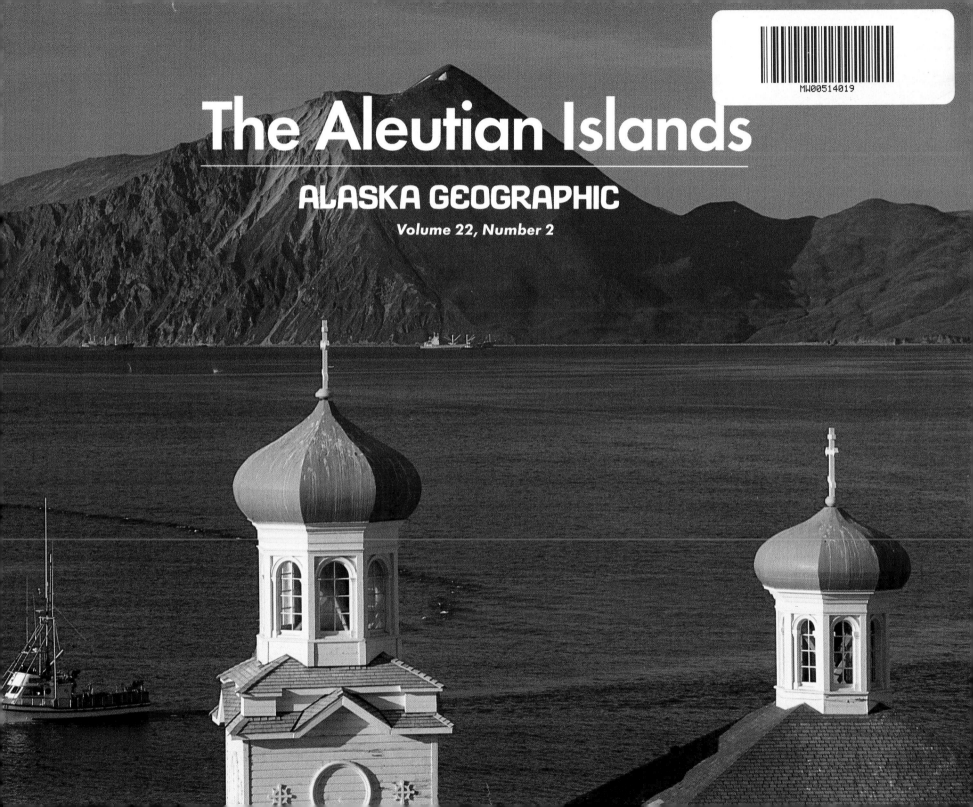

The Aleutian Islands

ALASKA GEOGRAPHIC

Volume 22, Number 2

To teach many more to better know and more wisely use our natural resources

EDITOR
Penny Rennick

PRODUCTION DIRECTOR
Kathy Doogan

STAFF WRITER
L.J. Campbell

BOOKKEEPER/DATABASE MANAGER
Vickie Staples

MARKETING MANAGER
Pattey Parker

BOARD OF DIRECTORS
Richard Carlson
Kathy Doogan
Penny Rennick

Robert A. Henning, **PRESIDENT EMERITUS**

POSTMASTER: Send address changes to
ALASKA GEOGRAPHIC®
P.O. Box 93370
Anchorage, Alaska 99509-3370

PRINTED IN U.S.A.

ISBN: 1-56661-026-5

Price to non-members this issue: $19.95

COVER: *Carlisle Volcano towers over Carlisle Island in this view from Chuginadak Island, both in the Islands of Four Mountains group.* (Nina Faust)

PREVIOUS PAGE: *Russian Orthodoxy represented by the onion domes of the Unalaska church and commercial fishing highlight key elements of the Aleutian lifestyle.* (Alissa Crandall)

FACING PAGE: *Most of the world's population of emperor geese winter in the Aleutian Islands.* (Cary Anderson)

ALASKA GEOGRAPHIC® (ISSN 0361-1353) is published quarterly by The Alaska Geographic Society, 639 West International Airport Road, Unit 38, Anchorage, AK 99518. Second-class postage paid at Anchorage, Alaska, and additional mailing offices. Copyright © 1995 by The Alaska Geographic Society. All rights reserved. Registered trademark: Alaska Geographic, ISSN 0361-1353; Key title Alaska Geographic.

THE ALASKA GEOGRAPHIC SOCIETY is a non-profit, educational organization dedicated to improving geographic understanding of Alaska and the North, putting geography back in the classroom and exploring new methods of teaching and learning.

MEMBERS receive *ALASKA GEOGRAPHIC*®, a quality publication that devotes each quarterly issue to monographic in-depth coverage of a northern geographic region or resource-oriented subject.

MEMBERSHIP in The Alaska Geographic Society costs $39 per year, $49 to non-U.S. addresses. ($31.20 of the membership fee is for a one-year subscription to *ALASKA GEOGRAPHIC*®.) Order from The Alaska Geographic Society, Box 93370, Anchorage, AK 99509-3370; phone (907) 562-0164, fax (907) 562-0479.

SUBMITTING PHOTOGRAPHS: Please write for a list of upcoming topics or other specific photo needs and a copy of our editorial guidelines. We cannot be responsible for unsolicited submissions. Submissions not accompanied by sufficient postage for return by certified mail will be returned by regular mail.

CHANGE OF ADDRESS: The post office does not automatically forward *ALASKA GEOGRAPHIC*® when you move. To ensure continuous service, please notify us six weeks before moving. Send your new address and your membership number or a mailing label from a recent *ALASKA GEOGRAPHIC*® to: The Alaska Geographic Society, P.O. Box 93370, Anchorage, AK 99509-3370.

COLOR SEPARATIONS:
Graphic Chromatics

PRINTED BY:
The Hart Press

The Library of Congress has cataloged this serial publication as follows:

Alaska Geographic. v.1-
    [Anchorage, Alaska Geographic Society] 1972-
    v. ill. (part col.). 23 x 31 cm.
    Quarterly
    Official publication of The Alaska Geographic Society.
    Key title: Alaska geographic, ISSN 0361-1353.

    1. Alaska—Description and travel—1959-
    —Periodicals. I. Alaska Geographic Society.

F901.A266   917.98'04'505   72-92087

Library of Congress 75[79112] MARC-S

ABOUT THIS ISSUE: After more than three decades with the U.S. Fish and Wildlife Service, much of that time spent in the Aleutians, Edgar P. Bailey retired in 1995 to his home in Homer, where he and his partner, Nina Faust, wrote the detailed account of the natural history of the Aleutians. Another Alaskan with long exposure to the Aleutians Janice Reeve Ogle wrote a personal account of some of the plants found in the island chain. Anchorage resident and experienced science writer Richard P. Emanuel provided the chapter on Robert "Sea Otter" Jones, one of the legendary figures in Aleutian lore. Charlie Ess, another Alaskan and a free-lance writer, detailed the life of Borge Larson, a fixture at False Pass for decades. Finally, accounts of the people, history, economy and communities of the Aleutians were written by staff writer, L.J. Campbell.

Many current and former island residents provided information, and we thank: John Concilus of Atka; Gilda Shellikoff and Mark Massion of False Pass; Emil Berikoff, Scott Diener and Colleen Petre of Unalaska; John Stamm, Simeon Pletnikoff, Val Dushkin and Agrafina Kerr of Nikolski; Daryl Pelkey, Jacob Stepetin and Joe Berenskin of Akutan; Steve McGlashan, formerly of Akutan and now of Anchorage; Dr. Douglas W. Veltre of the University of Alaska Anchorage; Father Paul Merculief, Russian Orthodox priest for the Aleutians; and Larry Cotter of Pacific Associates, Juneau, who works with the Aleutian Pribilof Island Community Development Association.

Contents

The Aleutians
Tiny Islands in Turbulent Seas

By Nina Faust and Edgar Bailey

EDITOR'S NOTE: *Early in 1995 Edgar Bailey retired from the U.S. Fish and Wildlife Service after spending 26 years working in the Aleutians and other parts of coastal Alaska. Nina Faust, a retired high school mathematics teacher, volunteered on many island expeditions for nearly 20 years.*

The Aleutian Islands are an International Biosphere Reserve, and all but 16 of more than 200 named islands are included in the Alaska Maritime National Wildlife Refuge, created in 1980. Originally all of the Aleutians were designated a national wildlife refuge in 1913. The Alaska Maritime refuge encompasses most of the pre-existing Aleutian Islands refuge and various other islands and mainland cliffs in the Bering and Chukchi seas, most islands south of the Alaska Peninsula, and some islands in the Gulf of Alaska. Stretching more than 1,100 miles, making it the world's longest small-island archipelago, the fog-shrouded, windblown, rugged, treeless Aleutians are the peaks of a submarine mountain range that bridges two continents and separates the Bering Sea from the Pacific Ocean. The Aleutians have 57 volcanoes, 27 active and 13 above 5,000 feet in elevation. The highest is Shishaldin Volcano on Unimak Island, 9,372 feet above sea level or 32,472 feet above the ocean floor. Active volcanoes are still shaping many of the islands today, creating new habitat.

Although winter temperatures in the Aleutians are mild compared to the rest of Alaska, the climate is extremely harsh and the islands are known as the birthplace of storms. Average summer temperatures are around 45 degrees, while winter temperatures are about 30 degrees. Low pressure centers sweep across

Rugged Chagulak Island, 3,786 feet high, is one of only two sizable Aleutian islands that escaped the introduction of foxes. A remnant population of Aleutian Canada geese was found here in 1982. (Edgar Bailey)

the Aleutians into the rest of Alaska or southward into the Pacific Northwest. The chain averages 90 percent cloud cover with more than 200 days of measurable precipitation and frequently experiences high winds, poor visibility and ubiquitous fog. Precipitation averages about 50 inches annually.

Eight named island groups comprise the Aleutians: the Krenitzen, Fox, Islands of Four Mountains, Andreanof, Delarof, Rat, Semichi and Near islands. Unimak is the easternmost and largest island, and has flora and fauna similar to that of the Alaska Peninsula. The chain arcs farther south than any point in

FACING PAGE: *At its peak, the naval reservation on Adak had 5,000 people, making it the eighth largest city in Alaska and the biggest voting precinct in the Aleutians. (Lon E. Lauber)*

RIGHT: *One of the subjects most closely associated with the Aleutians is weather, usually bad weather. While temperatures are not as extreme as in other parts of the state, winds can make any outdoor activity in the islands memorable. Here Nancy Hall hikes through wind-blown snow on Adak. (Lon E. Lauber)*

Southeast Alaska, dipping below 52 degrees latitude at Nitrof Point on Amatignak Island in the Delarofs. Nitrof Point is approximately the same latitude as the Queen Charlotte Islands in British Columbia. Interestingly, because the International Dateline runs north to south at 180 degrees longitude through Amchitka Pass, the Aleutians include both the easternmost and westernmost points in the United States. The westernmost point is Amatignak Island at 179 degrees, 10 minutes west; the easternmost point is Pochnoi Point, the eastern tip of Semisopochnoi Island. From a geological viewpoint, the Aleutian archipelago also includes the Commander Islands near the tip of the Kamchatka Peninsula in Russia.

Attu, the island closest to Asia, shares some flora and fauna common to both Asia and the rest of the Aleutians and frequently experiences migrants from the Kamchatka Peninsula and elsewhere in eastern Asia. For this reason, Attu is a popular destination for a specialized bird tour group called Attours that comes to Attu specifically to look for unusual Asiatic migrants, such as Terek sandpiper, olive tree-pipit, smew, common greenshank, Far Eastern curlew, rufous-necked stint, common pochard and Siberian rubythroat. Although the only commercial lodging available in the Aleutians is at Dutch Harbor, visitors can take cruises through these islands.

The Aleutians and their cold, turbulent, nutrient-rich waters host more than 250 species of birds. The tremendous tidal flux through the passes between islands creates strong currents and upwellings, which enhance these highly productive waters. There are more nesting seabirds in the Aleutians than in the rest of the United States combined. The Aleutians are also important wintering habitat for emperor geese, eiders, harlequin ducks and other seaducks. In summer, tens of thousands of shearwaters that nest in the southern hemisphere spend the austral winter in the Aleutians, especially in passes between islands and other areas far offshore.

Buldir Island has one of only four breeding populations of red-legged kittiwakes in the world. (Edward Steele)

The story of the Aleutians, its people and its wildlife, is one of tremendous change. Most people think the Aleutians are virtually pristine and untouched by modern civilization. On the contrary, even though most of the islands currently are uninhabited, few of them escaped the ecological catastrophe caused by the release of exotic animals and by other impacts of humans, especially World War II.

Beginning in 1750 on Attu Island, the Russians introduced arctic foxes for fur farming from the Commander Islands, where they are native. Mainly from 1900 to 1930 arctic or red foxes were released on more than 450 islands from the western Aleutians to the Alexander Archipelago in Southeast Alaska, with other fox introductions coming from the Pribilof Islands, the Alaska Peninsula, and as far away as Greenland. Nearly all islands in the Aleutians were used as fox farms. Most of the easternmost Aleutians, from Unimak to Umnak, actually have native foxes because the foxes could reach these islands from the Alaska Peninsula across ice in the winter or from former land bridges during the last ice age. When Vitus Bering first arrived in these islands, he found indigenous red foxes and thus named them the Fox Islands.

The effects of introduced foxes were devastating because ground-nesting birds evolved on islands without any terrestrial predators. The first indications that foxes were destroying phenomenally rich seabird populations came in 1811 when Natives on Attu and Atka islands reportedly had to go to nearby fox-free islands to obtain birds for their clothing and had to resort to using fish skins because bird skins were becoming difficult to collect. Fox farming escalated in the 1800s under the direction of the Russian American Co., but the peak came in the 1920s and early 1930s

when fur production was the third largest industry in Alaska, surpassed only by fishing and mining. Ironically, fur farms were originally encouraged on the Aleutian Islands National Wildlife Refuge, and by 1921, 23 islands were leased and stocked, mostly with arctic foxes. Fox farming thrived until the Great Depression, after which the industry never recovered.

Nobody really knew how destructive fox farming was in the Aleutians until the first thorough biological reconnaissance was conducted in 1936 and 1937 by Olaus Murie. He documented major marine bird and

Weathered relics of World War II clutter the shore, while tiny Loaf Island, connected to the Attu Island mainland at low tide, guards a corner of Massacre Bay. Since the 1940s the Coast Guard has operated a LORAN station at Attu. (Penny Rennick)

mammal rookeries, assessed the devastation of fox farming, and recommended that foxes be removed from some islands. He even correctly postulated that the foxes were responsible for the demise of Aleutian Canada geese.

A comparison of Murie's descriptions of seabird populations with more recent surveys shows an almost complete loss of birds on numerous fox islands. Except for birds that normally nest on cliff ledges or in rock crevices that foxes usually cannot reach, seabird colonies on islands inhabited by foxes are generally mere relics of populations that were undoubtedly once much more diverse, widespread and numerous. Since foxes cache food, they catch tremendous numbers of birds. Murie found a fox cache in 1936 that contained more than 100 auklets and he estimated foxes killed 40,000 birds a year on some islands. A recent study on a small seabird island in the Atlantic Ocean revealed that a dozen red foxes killed 31,000 seabirds in one summer. The annual damage of introduced foxes and other animals to seabird populations on islands in Alaska far exceeds the damage caused by a single event like the *Exxon Valdez* oil spill. Worldwide, introduction of exotic species has caused more extinctions of species and severe population reductions than any other factor. More than 70 percent of extinctions of bird species worldwide occurred on oceanic islands and were due to alien species such as cats and rats. For example, on the Kerguelan Islands in the Indian Ocean, rats and cats reportedly kill 1,200,000 seabirds a year.

The severe reductions of seabird populations on fox-inhabited islands throughout the Aleutians have impoverished this ecosystem. Bird excrement enhances vegetation by enriching the soil, creating the luxuriant plant growth found on productive seabird islands. Seabird islands become green much faster in spring than non-seabird islands. Seabird guano also locally enriches waters surrounding islands because the droppings have phosphates, nitrates and other nutrients used by phytoplankton that, in turn, benefit organisms

FACING PAGE: *Huddling against the wind, this tiny grove of Sitka spruce represents Adak National Forest. These trees were planted in 1944; in the two centuries since the Russians first settled in the islands, both they and Americans who followed have planted a few groves in this naturally treeless landscape. (Lon E. Lauber)*

RIGHT: *Fireweed brightens the shore of Unalaska Island's Captain's Bay in autumn. (Barbara Keller)*

LEFT: *Workers feed pollock into a filleting machine at a surimi plant at Dutch Harbor. (Harry M. Walker)*

LOWER LEFT: *A wealth of fisheries surround the eastern Aleutians, and in good times provide for a robust economy. But it takes money to purchase a boat, such as this purse seiner, and not all islanders who would like to participate have access to the fisheries. (Douglas W. Veltre)*

farther up the food web. Soils with decaying bones, shells and other refuse in ancient Aleut middens nourish similar luxuriant plant growth, another indication that the impoverished fox-farmed islands once supported the more lush vegetation found on well-fertilized seabird islands.

In 1949 efforts to remove foxes to restore nesting habitat for the endangered Aleutian Canada goose began on Amchitka Island. Robert "Sea Otter" Jones, a former manager of the Aleutian Islands National Wildlife Refuge and renowned Aleutian explorer, and former Gov. Jay Hammond were among the first to remove foxes from islands in the Aleutians. To date foxes have been removed by refuge personnel from 20 islands; foxes remain on at least 26 islands. On some small islands, fox populations died out on their own. Removing foxes from islands is recognized as the most important management action the Alaska Maritime refuge can implement to restore island biodiversity.

West of the Fox Islands, the Aleutians have no indigenous land mammals. Besides foxes, other mammals were introduced to some of the Aleutians, particularly in the eastern portion of the archipelago. Akun Island, for example, has cattle and horses, while Hog, Tangik and Poa have European hares that were released in 1957. Umnak and Unalaska have reindeer, cattle, sheep and horses. Indigenous land mammals on Unimak, their westernmost range in Alaska, include

ground squirrels, weasels, wolverines, wolves, caribou and brown bears.

Introduced ungulates are locally causing range damage by excessive grazing and trampling, including on two islands in the central Aleutians, Adak and Atka. Most of the islands with ungulate damage are largely Native-owned, except Adak, site of a Navy base. Atka is one of the most disturbed and is the only inhabited island between Adak and Umnak. Arctic foxes were introduced there in 1790, while reindeer were put on

Visitors to the Aleutians can hike the backcountry free from the worry of meeting brown bears, except on Unimak Island where this bear was photographed passing by on the far side of the False Pass airstrip. (Gilda Shellikoff)

Atka in 1914. By 1982, the reindeer had multiplied to 2,000. Caribou were introduced to Adak in 1958 to provide hunting for Navy personnel stationed there with their families. With the planned base closing,

A team of goose-catchers head back to camp below after a day spent scrambling up the rugged slopes of Buldir Island searching for Aleutian Canada geese. One hiker carriers a backpack with a special crate that can carry six geese. (Edward Steele)

there will be little hunting of caribou. Without hunters or native predators, the caribou population will quickly increase and cause severe damage to the fragile vegetation, once again illustrating the folly of introducing exotic species. The caribou are scheduled to be removed, but one suggestion calls for moving some of them to comparatively pristine islands off the Alaska Peninsula.

The best examples of the prolific bird life that was probably typical of most islands in the Aleutians when Bering first arrived are Buldir and Chagulak islands. These islands are not unique topographically or climatologically. What makes them unique is that they are the only sizable islands that escaped the introductions of foxes, rats or other alien species.

Buldir is the most isolated of the Aleutians. Westernmost of the Rat Island group, Buldir is located about 60 miles east of Shemya and west of Kiska, respectively, practically in the middle of a 120-mile-wide pass. This island of less than 5,000 acres with sea cliffs terminating straight into the ocean around most of its shoreline escaped fox introductions because it has no protected coves or good landing beaches.

Essentially pristine, this remnant of the Aleutian ecosystem has more than 3.5 million breeding birds representing 32 different species. Of the estimated 4.9 million seabirds nesting in the Aleutians west of Unimak Island, incredibly Buldir hosts about 70 percent of them but has less than 1 percent of the land area, making it one of the most diverse seabird nesting colonies in North America. It has the largest Aleutian population of breeding storm-petrels, Aleutian Canada geese and glaucous-winged gulls. The densest breeding population of parasitic jaegers is found on its inland slopes. The largest populations of ancient murrelets, common and thick-billed murres, and horned and tufted puffins found in the central and western Aleutians nest here. One of only four breeding populations of red-legged kittiwakes in the world is on Buldir. The other three are on the Pribilofs, Bogoslof Island and in the Commander Islands. Buldir also has the largest population of least, parakeet, crested and whiskered auklets; one of five known northern fulmar

colonies; one of the highest densities of breeding peregrine falcons; and is the westernmost breeding site for bald eagles.

Seabirds occupy specialized nesting habitats on islands. Burrow-nesters, such as tufted puffins, Cassin's auklets, ancient murrelets, Leach's and fork-tailed storm-petrels use the island's soft soils to dig their

Until military downsizing began in the mid-1990s, the Air Force operated a major station on Shemya Island, about 1,600 miles west of Anchorage and halfway between Anchorage and Tokyo. The 9-square-mile island, nicknamed "The Rock," is home to Eareckson Air Force Station. The U.S. military first occupied the island on May 28, 1943, during the battle to retake neighboring Attu Island from the Japanese in World War II. (Cary Anderson)

LEFT: *The Aleutians before the introduction of foxes provided ideal habitat for seabirds, such as this tufted puffin. Even two centuries after fox introductions and, more recently, fox eradication programs, the chain still contains major seabird colonies, although not in the numbers estimated for the pre-fox era. (Edward Steele)*

LOWER LEFT: *One of the highlights of birding in the chain is spotting the tiny, elegant whiskered auklet, whose easternmost habitat is the Baby Islands in Akutan Pass. (Lon E. Lauber)*

burrows. Marauding glaucous-winged gulls are surface-nesters that make their crude grass nests in vegetation often interspersed with the cavities of burrow-nesters.

Cliff-nesters like red-legged and black-legged kittiwakes, pelagic and red-faced cormorants, northern fulmars, and common and thick-billed murres vie for suitable ledges on precipitous sea cliffs and are segregated by elevation and ledge-size preferences. For example, the smaller red-legged kittiwake prefers smaller ledges than other cliff-nesting species and is usually found lower on cliffs than murres.

Tremendous labyrinths of habitat for crevice-nesters are found beneath enormous rock jumbles on Buldir and certain other islands. Horned puffins; parakeet, whiskered, crested and least auklets; fork-tailed storm-petrels; and pigeon guillemots nest in spaces between rocks. Despite its rich variety of habitat for seabirds, Buldir has limited habitat for such species as ducks, loons, phalaropes or terns. Also found on Buldir and most of the Aleutians are passerines, such as gray-crowned rosy finch, song sparrow, Lapland longspur and winter wren. Above 1,200 feet snowy owls occur in limited numbers during some summers. Their primary prey seems to be horned puffins.

Because it is so far from other islands, Buldir lacks ptarmigans that are found in large numbers on fox-free islands and in remnant populations on some islands. In fact, many islands have a unique subspecies of ptarmigan. On some islands, in particular Agattu, ptarmigans were extirpated by alien foxes.

Besides different nesting habitats, diverse feeding habits also distinguish seabirds. Some, like puffins and cormorants, feed on fish, while others like auklets and storm-petrels are predominantly plankton-feeders. Seabirds are also classified by characteristic feeding locations based generally on water depth around islands, namely near-shore feeders, offshore feeders (one to two miles offshore) and pelagic feeders (50-plus miles offshore). Often huge rafts of kittiwakes, murres and other species congregate offshore beneath cliffs vibrant with swirling masses of seabirds.

A feeding frenzy of kittiwakes churns the water to froth as the birds plunge into schools of fish, commonly northern lampfish or walleye pollock. Seabirds either obtain food as surface feeders or shallow divers, like kittiwakes and terns; or as deep divers like cormorants, murres and puffins. Storm-petrels, Cassin's and rhinoceros auklets, and ancient murrelets are rarely seen around colonies because they return to their nests only at night.

A National Historic Landmark, the Church of the Holy Ascension, known as the "cathedral" of the Aleutians, welcomes Russian Orthodox followers to Unalaska. Built in 1894-1895, the church has three chapels. An earlier church at Unalaska, built in 1825, was for a time the seat of Father Ioann Veniaminov, later St. Innocent, who became head of the Orthodox Church in Russia in 1868. This building is considered one of the nation's 12 most endangered historic landmarks and needs major foundation and roof repair. The city, village corporation, local businesses and individuals have donated to the restoration effort and Aleut reparation funds are earmarked for it, but more will be needed. (Harry M. Walker)

A few seabirds do not nest in colonies, but instead are solitary nesters. The secretive Kittlitz's and marbled murrelets lay a single egg on the ground, often on bare rocky slopes sometimes many miles from the ocean. Away from the treeless Aleutians marbled murrelets often nest on large tree limbs. Not surprising, only a few nest sites have been found for either species. Both Kittlitz's and marbled murrelets feed close to shore, mainly on fish and crustaceans.

A seabird colony is rather like a 24-hour factory with two shifts—the daytime, diurnal shift and the nighttime, nocturnal shift. On Buldir during the day, the constant calls from gulls, kittiwakes, murres and other birds blend with the sounds of sea and wind and the occasional roar of the few remaining sea lions. Toward evening, huge, tight masses of crested auklets whirl overhead swooshing down over the colonies with a sound much like an express train. As darkness approaches, the daytime sounds quiet and there is a lull where just the sea and wind are heard with only occasional bird calls. But then, when full darkness finally arrives, an incredible range of eerie chirps, twitters and ghoulish laughter fills the air as the legions of nocturnal seabirds return from distant feeding grounds.

The evening sky seen through night-vision binoculars is filled with thousands of small, delicate-looking, forktailed storm-petrels fluttering around like so many large moths. The nocturnal flights of thousands of storm-petrels, whiskered auklets and Cassin's auklets are one of the natural wonders of

LEFT: *Twenty miles long, Agattu Island lies 30 miles southeast of Attu in the Near Island group at the western end of the Aleutians. Lacking rats and bald eagles and with its population of foxes removed, Agattu proved a suitable island for the reintroduction of Aleutian Canada geese. (Staff)*

FACING PAGE: *Rough seas pound the shore of Adak Island. Storm seas and rogue waves spawn legends in the Aleutians. Bill Page, who made several cruises in the islands from 1944 to 1947, recalls: "During the Explorer cruise in 1944 we were returning to Adak for supplies. The ship was sailing on an easterly course in the Bering Sea.... I was scheduled to go on watch at 1600 so was seated at the officers' mess table.... That is, I sat at the inboard side of the table near the midpoint and faced the starboard side of the hull. The mess attendant placed a dish of butter in the center of the table and approached to set my meal in front of me. We were hit by a tremendous sea on the port side which heeled the ship over so far that I thought it wouldn't come back to stability. Food, coffee, dishes and personnel were pitched to the starboard and everything was a mess. The most amazing thing was seeing the butter dish glued to the starboard wall of the ward room by the butter it contained." (Lon E. Lauber)*

Red king crab await handling at Unalaska. The crab industry fueled a boom here in the late 1970s, but when declining stocks curtailed the fishery, the town also felt the pinch. (Barbara Keller)

until finally the precocious chicks leave their burrows and tumble down to the sea. Most burrow- and crevice-nesting seabirds have only one chick, which spends a month or more in the nest before fledging. The early departure of chicks has saved ancient murrelet colonies from extirpation on some islands infested with ground squirrels and other alien rodents. Nocturnal activity at colonies is an adaptation to avoid daytime predation from peregrine falcons, eagles, ravens and gulls.

The whiskered auklet, unlike its more social diurnal cousins, tends to be secretive and nocturnal around its colony. It is not found east of the Baby Islands in the eastern Aleutians. These monogamous crevice-nesters are the most ornately decorated of the auklets nesting in the Aleutians with three white plumes splayed across the face surrounding the eye and a quaillike, forward-curling plume on the crest of the head. The whiskered auklet lays a single egg and shares nesting duties with its mate. A near-shore zooplankton feeder, its main food is copepods. As the second smallest alcid, only the least auklet is smaller, whiskered auklets, like other auklets and most other seabirds, are vulnerable to fox and rat predation.

Aleutian Canada geese were nearly extirpated by introduced foxes in the Aleutians and on many islands south of the Alaska Peninsula. These geese, smaller cousins of the common Canada goose but with a distinctive white ring at the base of the neck, once nested on islands from southwest of Kodiak to the Kuriles north of Japan. A remnant population was first discovered on Buldir in 1962 and for a long time was thought to be the only surviving nesting population. Small populations of about 125 and 100 geese were also found about 20 years later on fox-free Chagulak Island in the Islands of Four Mountains and Kaliktagik Island in the Semidis off the Alaska Peninsula, respectively.

Aleutian Canada geese arrive at Buldir the first week of May and stay until the first week of September when

many seabird islands. Beautiful black-and-white ancient murrelets chirp and twitter like so many crickets. In July, just a day or so after their two chicks have hatched, adult ancient murrelets persistently call to their young they migrate back to wintering grounds in California. They are particular about the choice of vegetation type for nest location, almost always choosing beach rye-umbel/fern communities that include red fescue, their main food source during incubation. Beach rye is preferred for nest sites because the dried hummocks block wind and rain, and the green shoots grow out to provide cover from predators.

Recovery efforts for the Aleutian Canada goose began with fox eradication, a program of captive propagation of geese for release on islands where foxes were removed, followed by efforts to translocate wild geese from Buldir, and finally with the acquisition and protection of key migration and wintering areas in combination with closing goose hunting in wintering areas. Early in the program, geese were reared in pens at the Patuxent Wildlife Research Center in Maryland and then were released on Amchitka. Biologists think that bald eagle predation was the main limiting factor that precluded successful reintroduction of geese on Amchitka, which has the densest population of bald eagles in the Aleutians and a large population of Norway rats, probably introduced during World War II. Geese were also released on the now fox-free Near Islands of Agattu and Nizki, which do not have bald eagles or rats. Between 1971 and 1982, 1,000 captive-reared birds were released on these three islands, but by the early 1980s there were still no reported nests.

The captive-rearing strategy was modified to pair a captive female with a wild male, which was then translocated with the captive geese. This finally succeeded in establishing the first nests on Agattu in 1984 and on Nizki in 1987. Foxes were removed from Agattu by 1969 and from Nizki by 1976. Before the foxes had been eradicated neither of these islands had any remaining native geese. As the native goose population on Buldir increased, a third strategy for relocation to other islands was employed. Wild geese were captured by biologists using long-poled fish nets, an arduous and difficult task. More recently, border collies have been used to locate the geese. The geese freeze when the dog is right over them. Participating in a goose roundup is an exciting, but physically challenging experience, as recalled in our diary:

JULY 26, 1991, BULDIR ISLAND:
Got up for an early breakfast and then went ashore on Buldir to help with the goose roundup. Left Buldir camp

Rob Lewis and his goose-finding assistants Kyle (left), Peat and Cap take a break from the rugged work of scouring tall grass on steep slopes of Buldir Island for skulking Aleutian Canada geese. (Edward Steele)

about 9:45 a.m. and hiked along a trail up into a valley that progressively narrowed into a lovely mountain stream. Higher up the valley the lush waist-high vegetation gave way to alpine tundra. We went over a 1,200-foot pass and then dropped down into Icicle Valley. We were beginning to wonder where all the geese were. This operation is much easier now because the drive uses border collies to round up the geese. Six people carried backpacks with special crates for holding geese. Each pack can carry six geese. The two sheep dogs were controlled with a special whistle and voice commands from their master. Almost everyone had a long-handled dipnet to catch the geese.

Finally we got into the main part of Icicle Valley where the vegetation is very lush. Here we ran into geese almost immediately. People get really excited when geese are spotted. Though flightless in late summer when molting their feathers, the geese can quickly run through the grass and hide. They are real skulkers. The dogs were good at locating them. Some of the geese we got were single adults, so they were sexed, banded, and blood samples were taken before release. We were trying to catch as many goslings from a brood as possible along with one of the parents. We

got 19 fairly quickly. It is very tiring hustling through the tall vegetation and irregular terrain.

After lunch we made a final effort to catch another batch. We stumbled on a bunch of goslings and within minutes of our mad scramble we had seven more goslings. That was about it for the day. Both people and dogs were tired.

False Pass, population 90, borders Isanotski Strait on the eastern shore of Unimak Island. Residents of this primarily Aleut fishing community depend on commercial fishing and subsistence to support their economy. The shallowness of the Bering Sea side of the waterway separating Unimak from the mainland gives the community its name. (Gilda Shellikoff)

Akutan Volcano dominates the island of Akutan. The 4,300-foot volcano last erupted from March through May of 1992. (Scott Darsney)

JULY 27, 1991, LITTLE KISKA:

Traveled [by boat] to Little Kiska overnight. Smooth ride. Got up at 7 a.m. for breakfast and by 8 were on our way to the beach with the geese from Buldir. All the geese were taken out of the crates and put into the backpack carrying cases. We hiked about a mile up the hill to a small lake where the geese were released. Everyone let their birds out pretty much simultaneously. They took off immediately into the water and swam across the lake. They were very vigorous and apparently in good health despite the stress of the transfer.

In summer 1994, biologists stopped counting goose nests on Agattu after 270 nests were found, and there were probably more than 300. The wintering population of Aleutian Canada geese is now about 15,000, most from Buldir, which is enough of a population increase to upgrade the status of the Aleutian Canada goose from "endangered" to "threatened." The Aleutian Canada goose recovery project has been a remarkable success. Other birds such as common eiders, black oystercatchers, cormorants and puffins, are also making a spectacular recovery on islands where foxes have been eradicated.

In 1982, the second surviving wild population of Aleutian Canada geese was discovered. For many years, biologists had suspected that of all the Aleutian Islands they had visited, rugged, ethereal, fog-cloaked Chagulak, which had not really ever been explored on land, was the most likely island to have another surviving population of geese because foxes never were released there. Olaus Murie also suspected Chagulak could have geese. This 3,786-foot-high island is topographically somewhat similar to Buldir and has essentially the same plant communities. Buldir and Chagulak have the largest and most diverse marine bird populations in the Aleutians, and Chagulak has one of the world's largest northern fulmar colonies. Chagulak is more rugged than Buldir and also has no suitable landing beaches or protected coves. Roughly circular with about 3 square miles of area, its formidable cliffs, numerous knife-edged ridges and rocky slopes create forbidding topography for humans. We were very excited about the opportunity to land on Chagulak and explore the island.

JUNE 14, 1982:

Ran over to Chagulak Island to do a skiff survey. Fantastic bird island! We could only see the lower 500 feet, but it was buzzing like a hornet's nest. Thousands of northern fulmars are packed in the grass. The slopes are just riddled with puffin and other bird burrows. This has to be one of the largest tufted puffin colonies in the state. Thousands of murres are stacked on the cliffs along with kittiwakes. A large gull colony goes way up into the clouds. Horned puffins are common in the rock crevices. Large flocks of crested auklets were amassing offshore around 8:30 p.m. We also saw least auklets and parakeet auklets. We landed

on the west side behind a tiny island. Water is available on this beach and a marginal campsite could be set up.

Chagulak is an incredibly steep, rugged island with a tremendous amount of nesting habitat in the irregular valleys and pinnacle slopes. Aleutian Canada geese surely

LEFT: This bald eagle has just been shot on Atka Island. Most islands in the eastern and central Aleutians have substantial bald eagle populations. (Harry M. Walker)

BELOW: Hot springs, boiling pots and hot pools characterize Geyser Creek on Umnak Island. The creek flows into Geyser Bight, an area of geothermal activity about 25 miles east of Nikolski. Several other geothermal sites are located nearer to population centers, and the possibility of using geothermal energy from Makushin Volcano to generate power for Unalaska and Dutch Harbor has been considered. Other hot springs sites located near Aleutian communities include Cape Kiguga and Andrew Bay on Adak, Summer Bay near Unalaska and Hot Springs Bay near Akutan. (Dee Randolph)

FACING PAGE: *Pumicestone Bay reaches deep into the narrow neck of Unalaska Island at the foot of the little-explored Shaler Mountains. (Dee Randolph)*

RIGHT: *Dan Prokopeuff skins a harbor seal at Atka. Aleuts have traditionally used marine mammals for food, clothing and household and hunting items such as skin boats. But scientists and Aleuts are concerned about severe declines in Steller sea lion and harbor seal populations for reasons as yet unknown. (Douglas W. Veltre)*

could be nesting here. We need to walk the slopes to find out. Despite its inhospitable terrain, a fox farmer once had a permit to use this island, but fortunately none were introduced.

We went out in the boat tonight. By day the cacophony of fulmars and gulls rings in the air. By night the din of storm-petrels surrounds the island. Occasional chirps of ancient murrelets punctuate the din along with the laughter of Leach's storm-petrels and calls of whiskered auklets. We floated along the southeast shore listening to the nocturnal chorus. Steller sea lions swam near the boat and dove under the water. The bioluminescence was quite a show; wherever the sea lions swam we could see their glowing trails through the water. Near their loafing spot, little glittering planktonic jewels danced on the surface. Currents swirling the kelp waved the phosphorescent plankton around. All in all, it was a surreal night.

TUESDAY, JUNE 15, 1982:

Calm, but fog playing around Chagulak's upper slopes. We hiked up a ridge on the east side into the densely packed fulmar colony. When fulmars are disturbed they vomit orange, oily, digested food in a 3-4 foot stream. Foul smelling stuff. We got a glimpse of the top of the island—a steep, craggy, ice-covered, massive hulk. On the southeast shore we hiked up a slope to about 1,000 feet where we

found goose scat! We finally, after hearing geese honking, saw a pair of elusive Aleutian Canada geese! We couldn't find a nest but they behaved like they were territorial. This was an exciting discovery. Hopefully we'll be able to find some nests tomorrow.

Our late night foray was a disaster. We went back to the little beach by the island and attempted a premature landing without looking it over carefully enough and ended up trapped in a nasty surge hole. We were banging around in the rocks and filling full of water as we tried to push the inflatable boat back out stern first. We kept getting hung up on a rock so we just pulled it ashore and unloaded it. We portaged out of there over to a little isthmus and launched on the other side of the beach. Returned to the ship. Cassin's auklets appeared more common in the lights a mile down the coast from where we anchored last night.

WEDNESDAY, JUNE 16, 1982:

The fog lifted about noon, long enough to go to the place we had been the night before. Easy landing. We hiked up a ridge that goes out to Chagulak's southwest point. We ran into a flock of 60 Canada geese! This is really a significant find since there are supposedly only about 2400 Aleutian Canada geese alive. If the weather holds up we'll go back up again tomorrow.

Myth and mystery surround Bogoslof Island, the volcanic outpost north of Umnak, that has risen from the Bering Sea, weathered away, and grown again within recorded history in the Aleutians. Because the island was never used for fox farming, its wildlife have maintained a more natural balance, even though marine mammals and nesting birds in the area at the time of volcanic eruptions have suffered. (Kevin Bell)

JULY 10, 1982:

Went ashore at Chagulak's north side. Five of us walked the area and finally near the end of our sweeping survey located an endangered Aleutian Canada goose nest! It had 4 eggs and one cold one that had been kicked out of the nest. Nesting on Chagulak has been confirmed!

Chagulak, like Buldir, is an incredible seabird island. It has 19 nesting species and more than 1 million birds. More than 80 percent of the nesting seabirds in the Islands of Four Mountains region nest on Chagulak. A colony in excess of 500,000 northern fulmars, nearly the entire population that nests in the Aleutians, represents the largest single concentration on any island in Alaska or more than 25 percent of the state's population. The Cassin's auklet population also appears to be the largest in the Aleutians and probably in Alaska, and Chagulak may have also the state's largest population of fork-tailed storm-petrels.

The discovery of geese on Chagulak hastened efforts to eradicate foxes from nearby islands to encourage natural recolonization when geese become too numerous for the limited amount of nesting habitat available on Chagulak. Young geese might pioneer nearby fox-free islands and establish new colonies.

In 1983 the foxes were removed from 12,500-acre Amukta, Chagulak's closest neighbor. When biologists worked on the island, they reported there were bare cinders still warm to the touch and that steam was rising on the north side. Amukta has a roughly 3,500-foot active volcano that has spewed massive lava flows and cinder across the island, resulting in considerably less vegetation cover. Bird populations remaining are mere relics of what Murie reported on his survey in 1936. He noted this island as an excellent example of foxes living on birds. He found huge numbers of fork-tailed storm-petrels in vegetated lava flows. After 60 years of predation only occasional storm-petrels were heard. Amukta was the third largest source of arctic fox pelts in Alaska. Records show 1,530 fox pelts worth $44,468 were taken from Amukta. By 1984 the foxes had been eliminated, and within a few years the first Aleutian Canada goose nest was reported here.

Other islands in the central Aleutians that recently have had foxes removed include Kasatochi, Yunaska, Kagamil, Herbert, Uliaga and Carlisle. Lying nearly 12 miles off the western tip of Atka Island, Kasatochi has a substantial crested auklet colony that foxes were thriving on. Foxes were so numerous that they were even using the inhospitable interior of the precipitous caldera in the island's center. In summer 1993 Yunaska was cleared of foxes, and the following summer, Aleutian Canada geese were translocated from Buldir to Yunaska.

Sometimes fox eradication leads to unexpected archaeological discoveries. On one island in the Islands of Four Mountains an ancient Aleut burial cave was discovered. Jeff Wraley, a Fish and Wildlife Service seasonal employee and amateur archaeologist, found artifacts on a beach and then followed a trail of stone tool fragments and bone chips up steep talus to the eroded lip of a cave containing mummified bodies, artifacts, bidarka parts and bentwood boxes that are roughly a thousand years old.

Mummy caves have been found on other islands, but these had already been raided by early archaeological expeditions or by fox farmers. Foxes also damaged the mummies. Wraley reported seeing a fox gnawing on a mummy bone in the cave. Foxes regularly use burial caves for dens. Murie even noted foxes eating skin from Aleut mummies on Kagamil.

On Uliaga as well as on nearby Adugak Island, an experimental technique for removing arctic foxes was tried. Old fox farmers knew that red foxes would kill arctic foxes if they were on the same island. The late Henry Swanson, an old-time fox farmer from Unalaska,

For decades, Reeve Aleutian Airways has been the primary commercial air carrier for the chain. Here Capt. Thomas Hart (left), co-pilot John Minton and engineer Tim Kottre crew a flight between Anchorage and Shemya. (Cary Anderson)

said, "If you put blue (arctic) foxes on an island with red foxes, the blue foxes disappeared." Historical records show that Chuginadak in the Islands of Four Mountains had red foxes probably introduced by the Russians in the 1800s. In the 1930s arctic foxes were released there, but only the reds survived. In fact, Chuginadak with its 5,700-foot steaming Mount Cleveland, the highest volcano between Umnak and Great Sitkin, is the only island west of Umnak with red foxes.

In 1983 three male red foxes were captured in padded leg-hold traps near Nikolski on Umnak Island. They were released a few days later on Adugak Island with the assistance of Sergie Ermeloff from Nikolski. In May 1984 five pairs of sterile red foxes (vasectomized

A Markair Express plane lands at the community of Akutan on Akutan Island in the eastern Aleutians. (Alissa Crandall)

family that also includes shearwaters and petrels. In the Aleutians most fulmars superficially resemble dark-colored gulls, but they have large tubed nostrils atop their bills. Unlike noisy gulls and kittiwakes, fulmars are quiet, and even when flushed from the nest they are practically voiceless.

Murres will also nest on gentle slopes and sometimes atop flat islands if no foxes or other introduced mammals are present. Dense groups of both thick-billed and common murres pack grass-covered slopes and virtually every possible cliff ledge on Bogoslof Island near the Fox Islands. Bogoslof and Fire islands lie roughly 40 miles north of Umnak and have been volcanically active during the last 200 years. In May 1796, the volcanic island known as Old Bogoslof grew out of the ocean near lone Ship Rock, and in 1882 another eruption created nearby Fire Island. The first vegetation began to appear in 1817. Some eruptions have added new habitat for seabirds. The 1927 dome-shaped lava plug that was formed from a lava vent provided nesting areas for murres, kittiwakes and cormorants on its nearly vertical cliffs. This eruption again joined Bogoslof and Fire islands, but by 1935 erosion had once more separated the two. The most recent eruption in 1992 created a new dome that will provide additional breeding habitat for seabirds.

In 1909 Bogoslof was designated a sanctuary for sea lions and marine birds, four years before the rest of the Aleutians became a national wildlife refuge. Bogoslof escaped fox farming because it is very small and too unstable because of volcanic activity. Throughout the years, amazing changes in physiography, vegetation and avifauna have made Bogoslof perhaps the most dynamic island in the Aleutians.

Of course, the volcanic eruptions cost many animals their lives, and an eruption in 1883 reportedly destroyed many sea lions. Erosion has reduced nesting habitat tremendously as rock cliffs crumble to boulders

males) from Umnak were released on Uliaga. After several years, only red foxes were found on both islands. Although not practical for large islands because too many red foxes would be needed, this biological control technique works on small islands. Scientists hope the efforts to create fox-free habitat on islands near Chagulak will increase the small Aleutian Canada goose population in the eastern Aleutians.

The northern fulmars nesting on Chagulak show the adaptation of a cliff-nester to also use steep slopes with vegetation. Most of the fulmars on Chagulak nest on steep vegetated slopes, whereas on most other islands they use cliffs. Northern fulmars are fascinating birds that are classified as tube-nosed swimmers, the same that in turn are pounded by the sea into sand. Hence, population densities on Bogoslof have varied tremendously throughout the years as habitat was created and destroyed.

Bogoslof supports a wide diversity of seabirds and marine mammals. Both red-legged and black-legged kittiwakes nest here as well as red-faced and pelagic cormorants, glaucous-winged gulls, tufted and horned puffins, pigeon guillemots, common and thick-billed murres, storm-petrels and bald eagles. Bogoslof has a significant sea lion rookery, but its numbers have crashed from 3,300 in the late 1970s to just 540 in 1992. Northern fur seals commonly occur there. They also are regularly sighted at Ugamak Island in Unimak Pass.

The marked decline of Steller sea lions is widespread throughout the Aleutians and much of the rest of Alaska. Sea lions used to be the most abundant pinniped in the chain. They prefer offshore rocks, small islets and exposed points, and sandy or rocky beaches to haul out. They breed on isolated islands, such as Agattu, Seguam, Ulak, Buldir, Bogoslof and Ugamak.

Overall populations have decreased as much as 80 percent in recent years. Since 1960 the eastern North

The three main peaks of Unimak Island rise like sentinels in this view from the Alaska Peninsula. From left are Roundtop, 6,140 feet; Isanotski Peaks, 8,025 feet; and Shishaldin, 9,372 feet. Shishaldin is the highest peak in the Aleutian chain. (John Sarvis)

Pacific population has dropped from about 140,000 sea lions to 25,000. Ugamak was once considered the largest sea lion colony in Alaska and the world. Censuses revealed more than 13,500 animals in 1968 but only 450 in 1989 and 950 in 1992. In 1962, Sea Otter Jones reported a sea lion population on Buldir of more than 10,000 covering all available beaches. He reported he had to "haul the dory high above the beach lest she be wrecked by their activities." The Buldir population is now a mere 450. Amak Island also used to be a huge rookery with more than 4,500 sea lions. Now only about 870 are there. No one yet has proven

For the thousands of military personnel that have been stationed on Adak in recent years, Mount Moffett, 3,924 feet, provides a perfect setting for downhill skiing. The mountain commemorates Rear Adm. William Adger Moffett (1869-1933), who perished with the USS Akron in the Atlantic on April 4, 1933. (Joe Meehan)

what is causing the drastic declines, but many scientists agree that young sea lions do not appear to be getting enough to eat and that probably overfishing is a direct cause of the declines along with natural cyclic changes.

Sea otters were once virtually extirpated from the Aleutians by excessive exploitation by Russians and other fur traders. They have since reinhabited these areas. In 1962, Jones made the first recorded sighting of a sea otter at Buldir; they had not been seen in that part of the Aleutians in more than 50 years. Now the greatest population in the world is thought to be in the central and western Aleutians. Harbor seals are also common in most of the Aleutians, but they too are declining sharply in many areas. Other commonly seen marine animals include killer whales, Dall porpoise, minke whales, Pacific harbor porpoise and beaked whales. Close monitoring of seabird and marine mammal populations is needed to track the depletion of fish stocks, gillnet mortality, pollution and other human disturbance as well as natural fluctuations due to climate and oceanographic factors. Seabirds and marine mammals are indicator species for overfishing and pollution.

Another serious threat to seabird populations is rat introductions. At least 22 Aleutian Islands have Norway rats. In 1780 rats invaded Rat Island from a wrecked Japanese schooner, and that is why Russians named this group the Rat Islands. When foxes are removed from islands with rats, as on Kiska, rat populations grow even larger. Rats destroy eggs and chicks as well as damage habitat, and they can reach nests in small crevices inaccessible to foxes. They also compete for vegetation and burrows, as do ground squirrels, another exotic species that adversely affects seabirds on some islands. Both rats and foxes on the same island are especially deleterious to birds, but just removing foxes does not lessen the threat of rats to nesting seabirds. Unfortunately, total extermination of rats is extremely difficult and probably not possible on large islands. To date the biggest island all rats have been eliminated from is a 450-acre island in New Zealand. Other alien rodents like mice and voles also prey on seabird eggs and chicks.

Despite introduced predators, several islands in the Aleutians still have awesome populations of seabirds nesting in rock crevices. Some colonies in talus are so large that they are able to swamp the foxes with their enormous numbers and relatively inaccessible nests and survive. Kiska's lava flows on Sirius Point host the world's largest crested and least auklet populations with more than 1.4 million birds. Sirius Point is considered one of the most important marine bird colonies in the Aleutians. Despite foxes, Gareloi Island in the Andreanof Islands is one of the finest seabird islands in the refuge. Foxes have decimated the surface- and burrow-nesters, but crevice-nesters like least and crested auklets are able to maintain breeding populations in fractured lava and steep talus slopes.

Archaeological surveys throughout the Aleutians have revealed large numbers of former Aleut village sites. When the Aleuts first arrived, most of the islands in the Aleutians were rich in marine mammals, birds and other resources. Shortly after foxes were introduced to numerous islands, declines in seabirds resulted in the loss of food and materials for clothing. The Aleut people suffered population declines, just like the seabirds and marine mammals they depended on. When the Russians first arrived there were an estimated 15,000 to 25,000 Aleuts. However, by 1831, disease and killings reduced their population to less than 1,000. Now in the Aleutians there are villages only on Atka, Umnak, Unalaska, Akutan and Unimak islands.

The rich archaeological heritage attests to the probability that there were once many islands like Buldir and Chagulak. With continued eradication of foxes, prevention of the spread of rats and other

Between Akutan and Unalaska islands lie Unalga Island and the Baby Islands. In an increasing effort to develop tourism in the Aleutians, marine excursions are now being offered from Dutch Harbor to the Baby Islands to search for seabirds and marine mammals. (Douglas W. Veltre)

rodents, and careful management of fisheries, the original phenomenal biodiversity of the Aleutian Islands can be largely restored. Aleutian Canada geese and other waterfowl will pioneer islands in their former range if they are also protected in their wintering areas. Passerines, ptarmigans, oystercatchers and other birds should again flourish on numerous islands. Biologists hope seabird populations on restored fox-free islands will once again swirl with all varieties of seabirds that will, in turn, enrich their verdant breeding habitat with recycled nutrients. Such was the vibrant ecosystem that sustained the resourceful Aleuts for thousands of years.

One Woman's Aleutian Bouquet

By Janice Reeve Ogle

Lupine (far right), wild geranium (center foreground), wild iris (rear) and other wildflowers collected from the Atka tundra make a memorable Aleutian bouquet. (Douglas W. Veltre)

EDITOR'S NOTE: *Janice is the daughter of the late Bob Reeve, pioneer Alaska pilot and founder of Reeve Aleutian Airways.*

Rhododendrons and orchids in the Aleutians? You bet. Primroses, too. The floral variety in the Aleutians is almost endless, and at the same time unique, because of a combination of climate, terrain and location. After frequenting the Aleutians for more than 30 years as a Reeve Aleutian Airways flight attendant, my breath is still taken away by the summer sights. Prior to publication of Eric Hulten's *Flora of Alaska and Neighboring Territories* in 1968, my wildflower identifications included a lot of guesswork and were often inconclusive. Since then, I've had few uncertain identifications.

Plant collection and study in the Aleutians, classified as Hulten's "5th floral region," took place as early as the late 1700s during Russian exploration. Hulten's study in 1932 is the most complete and most useful.

All environments are represented in the Aleutians: saltwater seacoast and beach, fresh and saline lagoon waters, lakes, creeks, rivers, swamps, lush meadows, alpine tundra and rocky cliffs. Aleutian winters are characterized by frequent and severe storms, but as luck would have it, storms are few in summer. In spite of the short growing season, wildflowers flourish in temperatures averaging 50 degrees. A midsummer spell of dense fog in late June seems only to encourage a spectacular display of flowers in July and August.

My personal favorites are the species occurring also on the Kamchatka Peninsula off Russia's eastern coast, just 400 miles from Attu Island. The tiny Kamchatka rhododendron (*Rhododendron camtschaticum* subsp. *camtschaticum*) inspires visions of the Pacific Northwest's showy varieties, but ends with a distinct preference for the Aleutian's modest, little ground-hugger. The tiny, brilliant, magenta petals remind me of the rich but little-known grandeur of the Aleutians. So does the Kamchatka thistle (*Cirsium kamtschaticum*) that is found only on the westernmost islands of Shemya, Agattu and Attu. This hardy Russian migrant causes me to look westward and consider the Kamchatka inhabitants also walking among them.

Another personal favorite is *Senecio pseudo-Arnica* – I've always called them Aleutian sunflowers – that decorate all sandy Aleutian beaches in late summer with a beautiful yellow lining. Also prevalent on and near the beaches is rye grass, the raw material of exquisite Aleut basketry.

Found all along southern coastal Alaska and in the Aleutians is the Kamchatka lily (*Fritillaria camschatsensis*). Hulten describes the color as purplish-black, but to me it is a rich brown, the only brown flower I am aware of. It has several nicknames: squaw lily, from the use of its "rice," (the bulbs that contain starch and sugar) by Aleut women, and stink lily, from my summers on Umnak Island as a child and referring to its less-than-lovely scent.

Aleutian orchids are particularly fragrant, especially the bog orchids (Plantanthera). Convallariaefolia, with their greenish flowers, grow knee-high in some locations. Other orchid species vary widely in size and shape. The key flower (*Dactylorhiza aristata*) is also in the orchid family and displays a vivid magenta much like the rhododendron.

Hardy Aleutian primroses are the tall *Primula tschuktschorum* and the tiny *Primula cuneifolia*, little more than 2 inches tall.

Landing at any Aleutian airport in July brings views of runways virtually lined with lupines (*Lupinus nootkatensis*), an impossible landscaping feat except as arranged by nature. The backdrop of additional purple represents additional acres of lupines. If any flower were chosen to signify the Aleutians, it would have to be the lupine.

The island chain has its own special version of paintbrush (*Castilejja unalaskensis*) with an intricate combination of yellow and greenish petals.

Fireweed, Alaska's fall staple, occurs in eight species, in a wide range of height and showiness.

Two varieties of cottongrass grow in the Aleutians, both *Eriophorum russeolum* subspecies. If collected within a two-week period following maturation to full bloom but prior to onset of seeding, and then hung to dry, they make perhaps the most beautiful winter bouquet I know of, and evoke pleasant memories of an Aleutian summer.

The Aleutian shield fern, *Polystichum aleuticum*, is the Aleutian's much publicized rarity and is presently known to occur naturally only on Adak Island. Originally discovered on Atka Island in 1932, the plant has not been seen there since. The fern was discovered on Adak in 1975 and has been classified as an endangered species. Aleutian shield ferns grow on Adak on narrow ledges at elevations between 1,400 and 2,000 feet on steep, mostly northeast-facing, mountain slopes. The fronds are a dull green and only about 6 inches tall. These lovely ferns face an uncertain future.

The rollcall of Aleutian wildflowers is lengthy: iris, coastal fleabane (aster), anemone, monkshood, buttercup, chrysanthemum, daisy, dandelion, Pearly Everlasting, gentian, wild geranium, Grass-of-Parnassus, valerian, beach bluebells, Jacob's ladder, twisted stalk, monkey flower, moss campion, pussytoe, several saxifrage, spring beauty, starflower, sundew, twinflower and violets.

Visitors to the Aleutians can walk a short distance, sit down, and have well more than a dozen different kinds of wildflowers within arm's reach. Few experiences can equal being surrounded by wildflowers on an Aleutian island; a visual Eden complete with the song of Lapland longspurs, nature's own background music. There is a special kind of freedom in the remoteness of the islands, and a special kind of relaxation in being surrounded by the beauty of Aleutian wildflowers.

CLOCKWISE, FROM TOP RIGHT:

The showy blossoms of Senecio pseudo-Arnica brighten an Adak beach. (Cary Anderson)

Fall brings its own decor to the tundra-covered headland on Amaknak Island overlooking Captain's Bay. (Barbara Keller)

The endangered Aleutian shield fern clings to scattered nooks on steep slopes of Adak Island. (Lon E. Lauber)

Robert "Sea Otter" Jones

By Richard P. Emanuel

EDITOR'S NOTE: *A free-lance writer and former hydrologist with the U.S. Geological Survey, Dick last wrote about the Alaska Peninsula for* ALASKA GEOGRAPHIC®.

When Bob Jones began working for the U.S. Fish and Wildlife Service in the Aleutian Islands in 1948, the agency didn't have a picture of the sea otters that populated the waters under their jurisdiction. In describing the situation, Jones is not speaking metaphorically. It is true that Fish and Wildlife lacked information about sea otters under their care. But what Jones means is that they lacked so much as a single photograph of a sea otter.

"That was one of the first instructions I was to follow, 'Get some sea otter pictures!'" Jones recalls. "I carried that out in the winter of 1949, on Amchitka. The first one I sent back to Washington got an immediate response: 'Send the negative without delay!'"

Thus began a long career studying the wildlife of the Aleutians, including 27 years based in Cold Bay where Jones managed the Aleutian Islands National Wildlife Refuge. The far-flung refuge is now the Aleutian Islands Unit of the Alaska Maritime National Wildlife Refuge. Today, retired and living in Eagle River, near Anchorage, Robert "Sea Otter" Jones is still sought out by the Fish and Wildlife Service and others for his hard-won familiarity with the Aleutian Islands.

Sea otters have been intimately associated with the Aleutians since prehistoric times. Their dense fur and the high prices it commanded in the Orient drew Russian hunters and traders to the Aleutians more than two centuries ago. Robert Jones' first assignment as the newly appointed manager of the Aleutian Islands National Wildlife Refuge was to photograph the sea otter, and it was his initial work on the biology of this species that earned him his nickname, "Sea Otter." (Lon E. Lauber)

Izembek Lagoon was the focus of much of Sea Otter Jones' later research. Here he scopes the lagoon's shore in this October 1974 photo. (Jim Rearden)

Jones' association with the Aleutians began more than a half-century ago, during World War II. Born and raised in South Dakota, he earned a biology degree from South Dakota State University, in Brookings. He was a licensed radio amateur when he was called to active duty in the Army in 1941, so he wound up in the Army's Signal Corps school, in New Jersey.

A thin man with a hawklike demeanor, Jones speaks with deliberate care. "I had conceived a desire to come to Alaska," he says, "so I made that known. There were no assignments in Alaska at the time, so I was sent to Fort Lawton in Seattle, that was the nearest opening."

In August 1942, Jones landed on Adak, in time to endure his first Alaska winter in tents.

"We were equipped with coal-burning stoves, but there wasn't much coal," he recalls. "Of course there is no coal native to the Aleutians, so we had to adapt our stoves to burn diesel fuel. And a coal-burning stove is not very well suited to that."

The men on Adak devised a system for dripping diesel fuel from copper tubing onto hot rocks placed inside their coal-stoves. "And it would smoke and produce soot," Jones says. "It was necessary to clear the soot out of the stack every day. The accepted method of doing that was to open the lid, throw in a glass of water, slam the lid, jump on top and stand there while it would go *whoosh* and blow the stack clear. Then you'd have to scurry around outside and put out the burning fragments of soot. Occasionally somebody would misjudge it and all the soot would blow into the interior of the tent."

Eventually, the Army supplied stoves designed to burn oil, Jones says. "But we had mastered the other system fairly effectively by then. The wind was something we never did master."

In the Aleutians, winter winds were the Army's most vigorous opponent, according to Jones. "It isn't an extremely wet climate but it's damp all the time and a little moisture in a 40- to 50-knot wind is enough to go a long ways. Some of the guys couldn't hack it and they'd lose their marbles. Well, sometimes in the middle of the night the tent would collapse, a pole would break or the whole damn thing would blow away. And this didn't happen just now and then, it was a routine sort of business."

Just as troops on Adak were moving into Quonset huts and other more permanent buildings, Jones was sent 200 miles west to encamp on Amchitka Island. It was early 1943. "I remember it distinctly because we had one hell of a snowstorm, it obliterated all sign of us," he recalls. "Our own aircraft came over looking for us and they were unable to find us. Of course, it melted in two or three days."

Jones' job on Amchitka was to monitor Japanese aircraft. "There were seven or eight of us, a small detachment," he says. "We were sent out with portable radar, the first portable radar the U.S. Army ever put in the field. We went to Bird Cape, at the west end of Amchitka, where we could look into Kiska harbor," some 50 miles northwest.

Despite difficult conditions, Jones liked Amchitka.

The archaeological record can confirm use by prehistoric people of marine species, such as sea otters, when their bones are found in middens, the landfills of the past. Here Bureau of Indian Affairs archaeologists Mark Luttrell, Bill Sheppard and Fred Harden excavate a small test pit on Amchitka. The open book is used to establish an exact soil color match, in this case of dark organic soil and mussel and sea urchin shells. (Courtesy of Mark Luttrell)

"Probably most of all, it was that there were lots of animals there. There were lots of waterfowl and of course sea otters and other marine mammals. I was especially interested in the emperor goose, I had never seen that goose before, coming from the Great Plains."

Fifty years ago, the waters around Amchitka were home to the only large concentration of sea otters left in North America. Years of hunting had left the animals endangered elsewhere. "We knew the presence of sea otters there was important," Jones says. "The military command was aware of that and wanted to protect them to the degree possible, so those of us who were interested found our way into that extra activity."

The idea was to minimize the unnecessary destruction of sea otters and other wildlife. "That in itself was a fair undertaking because everybody was carrying

ABOVE: *Sea Otter Jones spent up to five months a year conducting field work in the Aleutians. With Jones at the Atomic Energy Commission base on Amchitka Island is the sea otter "Hortense." Amchitka was the focus of much of Jones' early studies of sea otters. Hortense, here rubbing her nose, was the first and youngest of the sea otters captured on Amchitka. She was a house pet of the Fish and Wildlife Service group until her mischievous antics forced her removal to the "otter house" on the pier. (U.S. Navy photo; courtesy of Robert Jones)*

TOP RIGHT: *In this 1950 photo, Capt. Floyd A. Puckett (left), commanding officer at Amchitka Air Force Base, and Sea Otter Jones chat beside the U.S. Fish and Wildlife Service's Grumman Widgeon. (Photo by E.P. Haddon; courtesy of U.S. Fish and Wildlife Service)*

weapons and they wanted to use them," Jones says.

Aerial target practice for the Army Air Corps posed another problem. "Off-lying rocks were used and there were apt to be sea otters around those. It was pretty much a case of trying to select rock outcrops and such that were remote from Amchitka. How effective that was, I never knew."

As the war continued, the Amchitka detachment grew and received more powerful radar. Japanese planes occasionally flew over the island, "but they never saw fit to come over and put us out of business," Jones recalls. "Of course, they had set up radar in Kiska Harbor themselves. We noted when their gear went off the air."

Jones and his colleagues speculated, half-jokingly, that the Japanese had left Kiska, a guess which later proved to be true.

During the course of the war, Jones served on several Aleutian Islands in addition to Amchitka and Adak. He was stationed on Ogliuga, east of Amchitka, and briefly on Little Sitkin Island and Tanaga. "So I saw quite a number of [islands]" he says, "much more so than most GIs."

Jones liked what he saw and elected to stay in Alaska after the war. He was in Kodiak when he got wind of a U.S. Fish and Wildlife Service opening. The agency had overseen a wildlife refuge in the Aleutians since 1913, but had never stationed an officer in the refuge. Now they wanted someone in Cold Bay. "I was able to land that job," Jones says. He began his career with the Fish and Wildlife Service in 1948.

"In my opinion, it was the finest assignment the Service had," Jones says of his posting to Cold Bay. "I wanted to be where there were animals and not many people, and it fulfilled both categories."

"My boss was a man of relatively few words when it came to outlining what I needed to do," says Jones. "He said, 'We are supposed to know all there is to know about the birds and other animals on the national wildlife refuge.' And that consisted of my instruction."

Jones threw himself into his job, focusing on a number of issues.

"We were trying to get a handle on the sea otters," he recalls. "The otters at Amchitka were observed to experience heavy mortality in winter. We were interested in the emperor geese, because that's where they winter, in the Aleutians." Near Cold Bay, at Izembek Lagoon, "we were interested in the populations of Canada geese and black brant and various ducks – pintails, teal, mallards. And there are swans there. We were just trying to pick up the threads of what the populations on a major wildlife refuge were doing."

Some information was already available in the scientific literature. Pioneering biologist Olaus Murie had ventured into the Aleutians before the war and had published important findings. Jones set about compiling what was known but he also spent five months a year in the field, much of it on Amchitka.

Field work in the western Aleutians depended on dories such as the one shown here being launched from the beach of Buldir Island in 1963. Sea Otter Jones is in the bow, handling the oars. The boat was being loaded when Jones spotted a series of waves approaching the beach. "I shouted that everyone should either get in the boat or get up on the beach," Jones recalls. Three people were in the dory with Jones when the first wave hit, including one man whose leg is all that is visible in this photograph. Karl Kenyon had scrambled up the beach before turning to see how his colleagues were faring. He unlimbered his camera in time to capture the second wave's arrival. After the rather eventful launching of their dory, Jones and his colleagues proceeded to survey Buldir Island's newly rediscovered Aleutian Canada geese, a race of Canada geese thought possibly to be extinct. (Karl Kenyon, courtesy of Robert Jones)

Aleutian islands sometimes harbor varieties of species that are different from their mainland relatives. Such is the case with the song sparrow, which in the Aleutians grows larger than its mainland counterpart and seems to become increasingly larger moving from east to west through the island chain. (Lon E. Lauber)

After he had acquired the Service's first photographs of sea otters in the wild, he began to puzzle out why Amchitka's otters were dying.

"The mortality occurred in February and it was at first thought that a parasite or bacteria was responsible. We did studies to learn what the invertebrate food resources were [for the otters]. Others conducted studies of nutrient requirements and feeding. I did quite a little diving."

Jones was almost certainly the first person to scuba-dive extensively in Aleutian waters, beginning in 1951.

"I learned it from friends in Los Angeles," he says. "It became apparent that if we were to get information about what happened below the water surface, we'd better go down and take a look. So I learned it."

Jones dove alone, with just an assistant in the boat. "Which is not a good practice," he warns. "So I was extremely cautious and it was slow going."

He dove off Amchitka, where sea otters were common, and for comparison, off Adak, where there were no otters. "At Amchitka," he found, "there were about six rock oysters in a square meter at the bottom; whereas at Adak, there were over 100" in a square 6 inches on a side. "In Constantine Harbor [off Amchitka], I found clam shells littering the bottom but I never found one alive. At Adak, it was just the reverse," live clams were abundant, empty shells rare.

The life cycle of the green sea urchin provided the final clue to the die-off of Amchitka's sea otters. The green sea urchin is the major food source for Amchitka otters, particularly during winter.

"The green sea urchin spawns in December or January," Jones explains. "While they are reproducing, they are high in food value. They are filled with either sperm, in the case of males, or eggs, in the case of females. But there comes a point when all these reproductive products are discharged into the sea. There is little muscle in the sea urchin and little food value in what is left. The otters kept on eating sea urchins but they weren't getting any nutritional value and downhill they went." By February, the otters began to starve.

It took four or five years to establish why Amchitka's sea otters were dying. "This was normal mortality in a population that had depleted its food resource," Jones says.

The die-off was part of a natural cycle, and once attuned to it, the scientists spotted evidence of past cycles in the kitchen middens or refuse heaps of ancient Aleuts.

"We didn't dig in the middens because we weren't qualified archaeologists," Jones says. But he often observed middens sliced open by storm erosion.

"You could start from the bottom and see a layer that would be all sea urchin shells together with sea otter bones," Jones says. "Then there'd be a layer with few of either, then a layer with sea urchin and otter remains.

Bald eagles reach the westernmost limit of their range on Buldir Island in the western Aleutians. Scientists think that the dense population of this predator on Amchitka Island precluded the successful reintroduction of Aleutian Canada geese on the island. (Lon E. Lauber)

LEFT: *Not surprisingly, several species of marine mammals are found in Aleutian waters, including Dall porpoise, among the fastest swimmers of any marine mammal. This species is most commonly seen by man when swimming in the wake of a ship. (Edward Steele)*

LOWER LEFT: *The bane of island bird populations in Alaska, foxes have created havoc in the Aleutians where they have led directly to the extinction of some ground-nesting species on many islands. Red foxes are native to larger islands of the eastern Aleutians where they were able to cross from the Alaska mainland. But most of the damage was caused by introduced foxes, such as this blue-phase arctic fox, which were first brought to the islands by fox farmers in the 1750s. (Cary Anderson)*

They're quite obvious and you can confirm it up and down the chain, as far east as the Shumagin Islands."

Time and again, the voracious otters expanded in numbers until they depleted the sea urchins. Then the otters died off, and after a time, both populations rebounded.

Sea otters have made a dramatic recovery in the Aleutians and across western North America since the 1950s. When Jones first studied them, the Amchitka otters, with others in the Delarof Islands to the east, were North America's only large group. There were a few otters in the Aleutians east of Adak, some in Prince William Sound and scattered sea otters off California. Today, all of these populations have expanded and widely colonized adjacent waters. Their recovery has been largely natural, due to protection from hunting.

"We did introduce them at Attu on an experimental basis" in the mid-1950s, says Jones. It was during this time that he acquired his persistent nickname, "Sea Otter" Jones.

"The number of animals we released at Attu was well below the level where the population could sustain itself. I concluded it was far better to extend the necessary protection to otters and let them expand than to try to introduce them. When a sea otter population really begins to grow, it will swamp the survivors of any artificial introduction."

A different kind of protection was necessary to foster the recovery of seabirds throughout the Aleutians. Before World War II, Olaus Murie had recognized the devastation wrought by foxes throughout the Aleutians.

"There are lots of pelagic birds in the Aleutian Islands, including nocturnal forms," which stay in their nests in the daytime, Jones explains. "Petrels are common, as are auklets. And they always nest underground. Of course some that are not nocturnal, like puffins, dwell underground with their young also. Introduced foxes played hell with these burrowing birds and in some instances wiped them out wholly."

The Fish and Wildlife Service decided to eliminate foxes from some islands to try to restore the original ecology. Already on Amchitka to study otters, Jones determined to start there.

"It took us a long time on Amchitka because we made a lot of false starts," Jones says. "And because of all the islands in the Aleutians, Amchitka was the best producer of foxes."

Jones and his colleagues trapped and hunted foxes to clear them from small islands. "On big ones we used poison," he says. "It's not legal to use poison now so it's been necessary to resort to traps. That makes it a difficult job."

One bird to benefit from the fox's demise is the Aleutian Canada goose. Once thought extinct, Jones and an assistant, Vern Berns, found 300 to 900 Aleutian Canada geese nesting on Buldir Island in 1963. The birds were tracked to wintering grounds in northern California, where they were given protection, Jones says. "Now there are 6,000 of them on Buldir Island and at least 100 nesting pairs on adjacent Agattu."

One of the Rat Island group, Amchitka Island is 35 miles long and about 3 miles wide. The island experienced three nuclear detonations between 1965 and 1971. (Staff)

Agattu had been the best producer of geese in the Aleutians, according to Jones, until foxes completely wiped them out. Now that the foxes are gone, translocated geese from Buldir Island as well as nocturnal birds are recolonizing Agattu.

"The last fox on Agattu that I personally killed," Jones says, "was a vixen carrying a fork-tailed [storm-] petrel in its mouth. So we were in time on Agattu, but just barely."

"That, in a nutshell, was what we did in the Aleutians," says Jones, "try to understand what the original ecology was and to restore it within the

possibilities. You can't do it completely but you can come a long ways."

In his effort to understand and safeguard the ecology of the Aleutian Islands, Sea Otter Jones ranged back and forth across the length and breadth of the wildlife refuge. Light planes sufficed for travel in the refuge's eastern end, from Cold Bay to Unimak Island, about as far west as you could fly and return in a day. Beyond that, light plane flight is folly. "If you attempt it, you are really missing something upstairs," according to Jones.

A frequent companion of Jones' in the eastern refuge was Fish and Wildlife Service agent and pilot Jay Hammond. In his 1994 book, *Tales of Alaska's Bush Rat Governor*, the former state chief executive recounts a couple of adventures he shared with Sea Otter Jones in the 1950s.

While landing a Supercub on a frozen pond in King Cove with Jones, Hammond leaped out to try to slow the plane with his feet. For his heroics, Hammond fractured both ankles.

"We had to replace the skylight in the cockpit and patch a crease in the wing before he could fly it again," Jones recalls.

On the hop back to Cold Bay for medical help, Hammond was unable to use his feet to control the

rudder-pedals. So Jones manned the pedals – from the back seat.

"Jay is a big man, he was sitting in the front seat of course, so I couldn't see much," Jones recalls. Hammond shouted instructions over the roar of the engine – left, LEFT! RIGHT!

"There was a bit of lag so our takeoff was not a model of perfection, but we did get it flying," laughs Jones.

The wobbly take-off made both men silently wonder how the landing would go.

"Then I thought, 'Well, to hell with it, there's no use sweating it,'" Jones recalls. He enjoyed the 15-minute flight then followed Hammond's commands as they lined up to bring the plane down.

"We wound up on the runway at Cold Bay and made a couple of loops down the runway," Jones says, "but

FACING PAGE: West of Adak Island, Kanaga Volcano, 4,287 feet high, dominates northern Kanaga Island and this view of Adak Island's Shagak Bay at low tide. (Lon E. Lauber)

RIGHT: World War II brought the Aleutians more prominence and turmoil than they had ever experienced. To counteract an unchecked Japanese advance along the island chain, the United States military built numerous installations, big and small, on island after island. These Quonset huts and buildings are sinking into the vegetation below 5,925-foot Tanaga Volcano on Tanaga Island. (Lon E. Lauber)

we made it and everybody lived happily ever after."

For field work in the western Aleutians, beyond Supercub range from Cold Bay, Jones relied on boats. He took commercial or military planes to bases at Adak, Amchitka or Shemya. At each base, he kept a 20-foot dory. A similar boat today sits on a trailer outside Jones' Eagle River home. A wooden model of a dory awaits completion in his den.

"They are about the most seaworthy small craft that has been devised," Jones says, "but that's for the open sea, she's an awkward boat in a river. And she's not fast."

A dory equipped with outboard motor and loaded with fuel, food and equipment would make 6 to 8 knots, says Jones. "Crossing Amchitka Pass, which is 60 miles, we were underway for 10 hours."

In calm waters, Sea Otter Jones brings the dory Water Ouzel ashore on Adak, circa 1955. (U.S. Fish and Wildlife Service, courtesy of Robert Jones)

The flat-bottomed dories "have some characteristics which make many people quite uneasy in them," Jones admits. "They will not stand still, the least little ripple will make the boat roll. But that's their saving grace, they will not shoulder up to the sea, they will roll away from it."

It's a saving grace which may have saved Jones' life more than once.

"We often found ourselves in rough seas," he says. "I would sail in 30 to 40 knots of wind if I felt the objective justified it. But we found ourselves in 60 or 70 knots occasionally, and it required an able sea-boat and a sailor who understood just what she could do and could not do."

Any close calls? "I suppose there were many," is all Jones will say. "I can't honestly say I was happy when it was blowing 60 knots but that happened to me on numerous occasions and I got to taking it in my stride. I much preferred not to get caught in winds like that because if you lose power, you've got problems."

As a means of self-defense, Jones early on learned to read the weather.

"Personally, I believe the Aleutians represent the easiest place to do that," he says. "It's because of the system of lows [low barometric pressure] marching from west to east through the chain in winter, one right after the other. The wind starts blowing from the southeast, swings to the southwest, to the northwest and then dies out. You'll have a nice day, you'll think, 'This is lovely.' But the day after, it's going to be southeast again. You know this if you pay any attention. It was worth our while to do so and we became quite good at it."

Cold, high-pressure systems from Siberia drift down over the eastern Aleutians from time to time. "Then you'll have calm weather and it'll freeze up," says Jones. "On one occasion, it was necessary to get an ice-breaker into Cold Bay. We had three feet of solid ice. That pressure system lasted for six weeks."

After months in the wilds, when Sea Otter Jones finally vacationed, he usually sought out the big-city culture and golden beaches of southern California, where his sister lived.

"One time when I went down there, my sister said, 'I have a woman I'd like you to meet,' Jones recalls. "She was a social worker at UCLA," a lively, intelligent gal with three kids from a previous marriage. They hit it off, and in 1963, Bob and Dorothy M. Jones were married. With her youngest son, Dorothy joined her husband in Cold Bay.

The one-room school in Cold Bay "and a backyard as big as however you'd want it" suited his stepson just fine, says Jones. "He went out into the Aleutians with me, not once but often."

Dorothy decided to return to school and enrolled in the University of California at Berkeley. For her doctoral dissertation, she did social and anthropological research among the Aleuts of Unalaska. Then she went to work for the University of Alaska's Institute of Social and Economic Research. Dorothy Jones has written many professional works and is now seeking a publisher for a novel she has written, set in the Aleutians.

Sea Otter Jones also spent a year doing graduate studies at the University of British Columbia, in Vancouver, Canada. He completed a master's degree at the University of Alaska in Fairbanks, in 1973.

Jones' graduate work focused on Izembek Lagoon, outside Cold Bay. "Izembek Lagoon has the largest single stand of eelgrass in the world," he says. "Mass is what makes that special. It produces over 4 million metric tons of organic material a year and that's all washed into the Bering Sea in the fall. You can break that down into carbon and phosphorous and all the elements that are critical to an ecosystem. Phosphorous is one of the more critical ones and there are some

Because the Aleutians and neighboring Commander Islands span from North America nearly to the Asian mainland, rarely seen marine creatures sometimes wash up on their shores. Here U.S. Fish and Wildlife personnel examine a Bering Sea or Stejneger's beaked whale that has washed ashore on Adak Island. (Lon E. Lauber)

1,500 tons of available phosphorous in the lagoon produced by eelgrass."

Eelgrass thus feeds the rich ecosystem of the Bering Sea, Jones explains. But first it feeds the vast numbers of waterfowl that gather and migrate through Izembek Lagoon. "Whether it's geese that feed directly on the leaves, pintails that feed directly on the seeds or the eider ducks that feed on the invertebrates resident on the eelgrass, it all comes back to that one thing."

Almost all the world's black brant, a small coastal

goose, migrate through Izembek Lagoon. The lagoon hosts one of the world's largest wintering populations of Steller's eiders, which nest mainly in Siberia, with small breeding populations in Alaska. Tens of thousands of Canada geese pass through the lagoon each year, as do emperor geese.

"I personally consider Izembek the best wildlife land the Fish and Wildlife Service owns," Jones says flatly.

Not that his proclamation is meant to slight the Aleutian Islands refuge.

"Each island possesses attractions, at least for me," says an expansive Jones. Plant and animal communities vary, sometimes from island to island. "Adak, for example, is entirely different from Amchitka," he says. "There's a dwarf tree, hardly more than a bush, the Asian mountain ash, that appears on Attu and

Agattu and that's all. That's characteristic of the Aleutians."

The islands are geologically interesting as well, says Jones. Some bear evidence of wave erosion on surfaces now well above sea level. Some are young, built from the sea by volcanoes within the last few thousand years. One winter day on Little Sitkin, Jones wandered among a field of steaming fumaroles.

"All the sand beaches in the Aleutians are black," Jones continues. They are composed of the mineral magnetite, eroded from volcanic rocks. "And because they are black, they absorb solar energy all out of proportion compared to the adjoining land. So the Aleutian beaches are much more productive than you'd expect, a lot of biological activity goes on."

Jones left Cold Bay for Anchorage in 1975. He retired from the U.S. Fish and Wildlife Service in 1980, after 33 years. Even in retirement, he continued to help on special projects in Prince William Sound, lower Cook Inlet and the Yukon River delta.

Despite these later contributions, it was the Aleutian Islands which, a half-century ago, captured the heart and soul of a young Army radarman from South Dakota. He left his own individual mark on the Aleutians, just as surely as the Aleutians marked him. And their mark is clear: They call him Sea Otter Jones.

LEFT: *Harbor seals, like Steller sea lions, have shown an alarming decrease in population in recent years. Scientists have not pinpointed the cause of these declines, but the general consensus is that it is related to insufficient nourishment for young seals and sea lions. (Lon E. Lauber)*

FACING PAGE: *Sea Otter Jones and Edgar Bailey, who collectively have tallied more than a half century of Aleutian experience, contemplate the relatively peaceful seas at Binnacle Bay on the north shore of Agattu Island. (Courtesy of Edgar Bailey)*

People of the Aleutians

By L.J. Campbell

Introduction

The Aleutian Islands sheltered some of the first people to come to Alaska some 4,000 to 8,000 years ago. They traveled the ocean in skin boats, fishing for halibut and hunting sea lions with bone-tipped spears. They dressed in bird pelts and seal-gut parkas, and lived in sod houses. They called themselves Unangan or Unangas, meaning "we the people."

Descendants of these ancient islanders still live here, in some of the nation's most geographically isolated communities. Many of today's Aleuts still rely on the sea, hunting and fishing for food, and occasionally speaking the language of their ancestors. Yet their traditions mesh with modern technology.

The five surviving Aleutian Island villages boast new frame houses, hydroelectric power, telephones, fax machines, satellite television and computers linked to the global information superhighway. The villagers haul their halibut to shore in motorized skiffs. They drive four-wheel all-terrain vehicles along beaches to gather kelp and mussels. They fly in floatplanes and jets to distant cities to shop in malls, have their babies in hospitals, lobby government officials and attend Native corporation meetings.

The Aleutian Islands have endured the curiosity of many nations. The islands' inhabitants experienced the earliest and some of the harshest contact with outsiders of any of Alaska's Natives. Foreign explorers, exploiters and settlers brought diseases and different rules, customs and values. Some of these, for better or worse, still mark Aleut society.

Three distinct periods in Aleut history can be easily

Kuluk Beach is the perfect site for a Fourth of July picnic on Adak. Military downsizing will substantially reduce the Adak population. According to the Alaska Department of Community and Regional Affairs, there were 5,017 year-round residents of the Aleutians in 1994. (Lon E. Lauber)

Young people in Atka started performing Aleut dance in 1995, in a dance revival project funded by the Aleutian Region School District. Traditional dance died out in the Aleutian Islands during Russian colonization, however remnants of Aleut dance are still performed today in the Commander Islands, where Aleuts were transplanted by the Russians in the 1700s. These dances were brought to Atka by Katya Gil, a folk dance expert and choreographer from Kamchatka in the Russian Far East. She lived in Atka nearly a year-and-a-half, recreating the dances and helping the youngsters sew traditionally styled garmets using sea otter and fur seal skins. Gil was aided by Shirley Hauck from Chicago, an anthropologist who studied descriptions of Aleut dance recorded in journals by early Russian explorers. Pictured here, left to right, are Annlillian Nevzoroff, Nina Golodoff, Louise Nevzoroff, Mary Swetzof, Katya Gil, Lucinda Nevzoroff (kneeling), Christina Golodoff, Larisa Prokopeuff and Nancy Zaochney. Another nine children who learned the dances are not shown. (John Concilus)

identified. The legacies of two blend with an evolution of the third to frame the Aleutians as we know them today.

The first major changes came with the Russians, who landed in the 1700s and made the islands stepping stones for Russian America. The Russians weren't welcomed, but they stayed more than 100 years. Their influences persist today, and the religion they introduced still dominates.

Even use of the term Aleut was a Russian innovation. Apparently the Russians adopted the word from islanders in the western Aleutians, and they applied it to all the other Natives they met.

The second significant period occurred during the 1940s, when the military descended. The Aleutians became a World War II combat zone between Japan and America. The only battles fought on North American soil took place here. The Aleutian campaign, overlooked by many war histories, was a bloody and sad chapter in American history. Among the casualties were lives of soldiers and civilians and chunks of Native culture, lost when Aleuts were banished to faraway, desolate camps. Now 50 years later, the Aleuts are receiving reparation money for wartime damages, and the government is trying to clean up old munitions and hazardous chemical dumps. Five National Historic Landmarks commemorate war fields in the Aleutians.

A third invasion makes its way through the islands today. Commercial fishing dominates the scene. Seafood processors, boat owners and fishermen from distant U.S. ports and foreign countries flock to the Aleutians for a coveted share of the world-famous Bering Sea and North Pacific catch. Their huge vessels haul tons of fish through towering ocean waves, while the village locals in their small boats watch from the sidelines and negotiate entrance into the lucrative deepsea fishery around them.

This activity makes Dutch Harbor, centrally located on Unalaska Island, the busiest fishing port in the nation. It is also a transportation hub for the chain, with daily jet service handling hordes of business travelers, fishermen and laborers. With this has come efforts to sell the Aleutians to tourists, and the islands, filled with exotic birds, wildflowers, scenery and history, are poised as one of the newest destination darlings in the adventure tourism business.

All the while, the people who live here year-round weather the Aleutians' mighty storms, both natural and manmade. They are finding a place in yet another scheme for their homelands imposed by outsiders, trying to save what remains of their culture and take control of their future while, at the same time, readying themselves for the next big blow.

Here's a closer look at the people of the Aleutian Islands.

"We, the People"

The first people to the Aleutians probably came from Asia near the end of the last great ice age, when low seas exposed a large land bridge between Asia and North America.

These early immigrants may have boated along the southerly coast of the land bridge, which reached almost to Umnak Island. Or they may have traveled inland, wandering down through the Alaska Peninsula. Perhaps they came from several directions in multiple migrations to create the maritime nation from which modern Aleuts descended.

In Unalaska, the boys of summer gather at 9 p.m. on the longest day of the year for a rousing game of baseball. (Alissa Crandall)

Archaeologists, anthropologists, linquists, other scholars and Aleuts have spent years trying to unravel this and other puzzles about the chain's earliest inhabitants.

Oleana Dirks of Atka learns how to make the finely woven baskets for which the Aleutian Islanders are famous. (Harry M. Walker)

Some 400 abandoned village sites are listed in state records, although two to three times that number are known to exist, spotted with binoculars from boats and airplanes or reported by Aleut hunters and trappers. Only a fraction of these sites have been scientifically studied, largely because of the cost of getting to their remote locations. Many others may never be found, having been eroded or submerged.

The oldest known Aleutian site, also the oldest

known site on Alaska's coast, is on Anangula Island, just north of Umnak. Artifacts from Anangula date back 8,000 to 8,500 years. Whether modern Aleuts descended from the early Anangula people or from a later-arriving people is unknown.

After Anangula, the archaeological records of the Aleutians lapse. The next oldest site dates back about 4,000 years. The Chaluka village site, near Nikolski on Umnak Island, is this age and shows continuous occupation from that time forward.

We have some ideas about early Aleuts, gleaned from Aleut stories, reports and sketches by early explorers and priests, artifacts from old village sites, and pieces of art, clothing and other cultural vestiges in museums and private collections.

These island dwellers naturally relied on the sea, spending most of their time hunting, fishing, gathering food and making clothing and tools. Driftwood that washed ashore and chunks of sod from the nearby tundra were basic building blocks. They lived near the ocean, on bays with gently sloped beaches good for landing boats. Fresh drinking water came from nearby springs or creeks. An ideal village site would be a neck of land stretching between two bays. If intruders attacked from one bay, the villagers could escape in their boats on the other.

Among their most important tools were superbly made boats, kayak-style, skin-covered bidarkas. A village, and an individual's, power and wealth related directly to the number of bidarkas. They were master seamen and from these ocean-going crafts, they hunted marine mammals, sea lions, seals, sea otters and even whales, as well as waterfowl. They also ate marine invertebrates such as sea urchins, clams, mussels and

Just like his ancestors, fisherman John Nevzoroff is at home on the sea. (Steve McCutcheon)

octopus; birds and eggs; ocean and freshwater fish; and for variety, plants and berries.

They made clothing and tools from animal bones, skins, intestines, stomachs, teeth and whiskers, along with bird feathers and beaks and dried plants. They fashioned rain gear from cleaned sea lion or seal intestines, split and sewn into dresslike garments. They sharpened bones into harpoon and spear points. They chipped stones, particularly basalt, into digging, cutting and hunting tools. They wove dried grasses into sleeping mats, socks and finely made baskets.

Before contact with outsiders, the Aleutian Islands harbored at least eight distinct social tribes from the different island groups. They spoke different dialects. Sometimes they traded with each other. They often fought each other over territories and women, enslaving their captured enemies.

When the Russians arrived, some of these groups had already disappeared. Even so, an estimated 12,000 to 15,000 Aleuts lived throughout the chain at the time of first, prolonged contact with outsiders.

Russian Era

Russians landed in the Aleutians in 1741 while exploring the Pacific coastline. Two Navy ships under Capt. Vitus Bering encountered Aleut warriors in bidarkas, in the Shumagins and near Adak Island. The Aleuts wanted the sailors' metal knives, disdaining trinkets and cloth. They were already familiar with iron, having obtained pieces washed ashore from shipwrecks or through aboriginal trade channels.

Fur pelts collected by this expedition spurred interest back home. Russian fur traders started flocking to the Aleutians. Some got along with the Aleuts, trading fairly for sea otter pelts and other valuables such as walrus tusks.

But most of the traders treated the Aleuts badly. The first Aleuts were murdered in 1747, when 15 warriors refused to surrender some metal for inspection. As the

LEFT: *Barabaras, semisubterranean sod structures, were standard housing for early Aleuts. As foreign goods penetrated their lifestyle, their traditional houses incorporated some of the items from the other cultures such as windows. Here Sergie and Agnes Sovoroff pose in front of a barabara at Sandy Beach near Nikolski in this 1972 photo. (Douglas W. Veltre)*

FACING PAGE: *Three-wheelers and four-wheelers have become the most common means for Aleutian Island villagers to get around. This Atka resident travels one of the main roads through the old village and past the Russian Orthodox church on the hill. (Lon E. Lauber)*

The Russians brought Russian Orthodoxy to the Aleutians in 1794 and it remains the dominant religion. Under the Aleut Restitution Act of 1988 and subsequent legislation, the federal government provided $4.7 million for restoration of Russian Orthodox churches in the region. Akutan has never had a resident Russian Orthodox priest, but the town's Chapel of St. Alexander Nevsky, built in 1918 to replace the original chapel, is due for restoration. (Steve McCutcheon)

together to kill shiploads of their unwanted visitors; the Russians retaliated, killing hundreds of warriors, attacking villages and destroying bidarkas, spears and other tools vital to survival. The Russian government disciplined some traders for mistreating the Aleuts, but complaints took so long to reach St. Petersburg that little could be done to curtail abuses.

During the late 1700s, Russian companies continued expanding deeper into Alaska. Other nations, including Spain, Japan, Sweden and England, sent explorers to claim lands in the North Pacific as well. Capt. James Cook, sailing for England, landed in the Aleutians at Unalaska Island in 1778. His reports of abundant sea otters and fur seals would bring British and American merchants into the region.

In 1794, the Russian Orthodox church sent priests to establish churches and schools. The clergy intervened on behalf of the Natives and helped preserve parts of Aleut culture during this time of dramatic transformations. Father Ioann Veniaminov, whose path to sainthood started in Unalaska, developed a written Aleut alphabet with the help of elders. He performed church services in Aleut, translated church literature, and his journals provide incomparable detail about Aleut life. Even so many Aleut stories and traditions, like dances, were lost. In some cases, traditions died with the elders because their heirs were away from the villages, occupied with sea otter hunting and other Russian activities.

In 1799, the Russian government granted a trading monopoly for Alaska to a company owned by powerful merchant and veteran trader Grigori Shelikhov. The new Russian American Co. continued forcing the Aleuts to hunt and relocated them to Russian settlements in Kodiak, the Pribilof Islands and Sitka in Southeast.

As time passed, the Aleuts gradually gained limited rights and freedoms. They took pride in their literacy,

fur trade peaked between 1760 and 1780, so did the violence, particularly in the larger, more populated eastern islands of Umnak, Unalaska and Unimak.

To control the Natives, the Russians often held Aleut children as hostages. In this way, they forced Aleut men to hunt sea otters. Repeatedly, Aleuts banded being schooled in both Aleut and Russian. A few privileged Aleuts attended Russian universities and academies, returning to the islands as teachers, mapmakers, shipbuilders, medical assistants, priests. Those employed by the company received wages and could advance as managers. Villagers sold their fur pelts according to an established payment schedule.

Then in 1867, Alaska became an American territory.

Early American Period

The change to American ownership did nothing to immediately improve the Aleuts' situation. They gained no citizenship rights and lost what little status they had acquired under Russian rule.

The Russian Orthodox Church continued its mission work, but getting money to operate was difficult. Numerous Russians, many of them married to Natives, remained in the islands and they, like the Aleuts, had to learn a new language and laws.

The Alaska Commercial Co. acquired Russian American Co. assets and conscripted Aleuts to hunt sea otters and fur seals. Under the Americans, the Aleuts had trouble obtaining enough skins for bidarkas, partly because the animals were becoming scarce. The Russians had limited the take of pelts to prevent overhunting, but the Americans had no such restrictions until the animals were almost extinct.

At first, the Americans knew little about the Aleutians. A Navy survey secretly conducted during the Russian era produced simple navigational charts. Subsequent government expeditions led by scientist W.H. Dall in the 1870s added information about the Natives, villages, plants and animals. He estimated about 1,384 Aleuts and 270 Aleut-Russians lived in the chain. Tuberculosis, malnutrition and alcohol abuse were noted as big problems.

Ships with the U.S. Revenue Cutter Service, later the U.S. Coast Guard, patrolled the seas. Their crews enforced American laws and served as doctors and sometimes social service agents. Steamships delivered supplies, mail and occasional tourists. Visits were infrequent until about 1891, when regular mail service began.

When Russians first arrived in the Aleutians, Aleuts lived in many villages scattered throughout the chain. As foreign influence spread, those villages consolidated. By the 20th century, there were many fewer villages, and World War II brought the consolidation and abandonment of even more. Twenty residents of Kashega on Unalaska Island were transported to southeastern Alaska on July 5, 1942. The village was abandoned after the war and many of the residents moved to Unalaska. (Douglas W. Veltre)

FACING PAGE: *Wild sheep, remnants of flocks brought to the island earlier in the 20th century, crowd this cliff on southern Unalaska Island. Various efforts at ranching have been made in the Aleutians with uneven success. When boats were still the standard means of traveling between islands, getting feed for cattle and sheep, supplying fresh meat to the boat crews and to fishing boats, and transporting the meat and wool out for processing were more economically feasible. In the mid-1990s, only Unalaska and Umnak islands held much prospect for ranching. Milt and Cora Holmes have operated a ranch at Chernofski on Unalaska Island for decades and ranchers from Canada and the Lower 48 are hoping to develop a viable operation on both islands. (Dee Randolph)*

RIGHT: *A fisherman picks salmon from a setnet at Unalaska. Fishing has been a subsistence activity among Aleutian Islanders for generations, but not until the American period did commercial fishing come to play an important role in the region's economy. (Alissa Crandall)*

Like any frontier, the Aleutians drew an assortment of people, most of them involved in trapping, fishing, ranching, whaling and mining.

Most of the islands were used during the early 1900s for fox farming. Cod, herring and salmon fishing brought fleets of boats, canneries and salteries to the region. Several ranchers imported sheep and cattle to the chain, adding to those brought earlier by the Russians. Chernofski on Unalaska Island and Nikolski on neighboring Umnak saw the most intensive efforts at sheep ranching, with a brief attempt on Kanaga Island. Efforts at cattle ranching were sporadic, and there was some mining, including a short-lived sulfur operation on Akun and gold mining on Unalaska.

Whaling in the Aleutians centered at a plant built by Norwegians in 1912 on Akutan. They hired local Aleuts in the plant and to crew whaling boats. The plant processed about 100 whales a year into oil, dog food and fertilizer, until 1942 when it closed during World War II.

World War II

U.S. military commanders started fortifying the Aleutians soon after war broke out in Europe. Japanese spies had been surveying the chain, only 650 miles from their home bases. Both countries recognized the islands as potentially strategic North Pacific holdings.

By 1940, the Aleut village of Unalaska swarmed with thousands of workers and troops as Fort Mears Army Base and Dutch Harbor Naval Station went in. On the island's opposite end, a dock was built at Chernofski Harbor to supply Fort Glenn, a secret airfield on Umnak Island.

In December 1941, Japan bombed Pearl Harbor, bringing the United States into the war. Activity in the Aleutians accelerated. Navy weather stations opened on Kiska and Kanaga islands. Before the buildup was

Peace has been restored to Unalaska this day in August 1942. Two months earlier the Japanese had bombed adjacent Dutch Harbor, beginning the Aleutian campaign that led to Japanese occupation of the islands of Attu and Kiska in the western Aleutians and the subsequent American counterattack that culminated in the battle of Attu in 1943. (Steve McCutcheon)

complete, military installations big and small were scattered throughout the chain.

On June 3 and 4, 1942, Japan bombed Dutch Harbor. This diversionary tactic was designed to draw U.S. ships away from their naval station at Midway in the central Pacific, the true object of Japan's desire.

The Dutch Harbor battle is detailed in *ALASKA GEOGRAPHIC®* Vol 18, No. 4, *Unalaska/Dutch Harbor* (1991). Briefly however, the June 3 attack on Dutch Harbor lasted about 20 minutes, before strong counterattacks turned away the enemy. Returning to their air carriers in the Pacific, the Japanese flew over Fort Glenn. A soldier hanging out hospital linens to dry spotted the planes; faulty radio connections had failed to alert Fort Glenn during the bombing. American fighters scrambled after the Japanese, shooting down two planes in a surprise attack.

The next day, the Japanese scored bigger hits on Dutch Harbor, bombing fuel tanks, a barracks ship, the power plant, a dock and hanger, gun emplacements and the Unalaska hospital. The enemy also discovered Fort Glenn and a dogfight ensued, bringing down two American planes and six Japanese bombers.

The two-day conflict cost 76 Americans lives, with 10 missing and at least 28 wounded.

Japan also wanted to occupy the western Aleutians, to block possible invasion of their country. Three days after raiding Dutch Harbor, about 1,250 enemy soldiers landed at Kiska. They captured the island's only inhabitants, 10 Americans at the naval weather station, although one escaped and survived 50 days before surrendering.

On June 8, another swarm of Japanese took Attu Island. As the soldiers came through Attu village, the government schoolteacher radioed Dutch Harbor from his house before being executed. The Japanese then captured the 42 Aleut villagers and the schoolteacher's wife, shipping them to prisoner camps in Japan.

With this, the U.S. military funneled its few available ships and planes in the North Pacific to the western Aleutians. The Navy moved seaplane tenders to Atka, to refuel and rearm fighters out of Dutch Harbor. Bombing runs buffeted Kiska. The Atka villagers were sent to their fish camps along shore and on the

night of June 12, Navy seamen torched most of Atka village including the church, a wartime tactic to erase landmarks.

Most of the villagers saw flames and tried returning home. They were loaded onto Navy boats, able to salvage nothing except what they wore. The next morning, the remaining villagers were picked up by seaplanes.

By now, the American leaders had decided to

Innokenty Golodoff, originally from Attu, and Mike Nevzoroff, from Atka, enjoy a game of cribbage. Innokenty, born in 1917, was one of the Attuans taken prisoner and sent to Japan when the Japanese invaded Attu Island, June 8, 1942. (Steve McCutcheon)

remove the Aleuts and non-essential civilians, ostensibly for their protection. Such a move had been discussed before but only after the bombing of Dutch Harbor, the capture of Attu and destruction of Atka was the plan implemented. It was poorly thought out, and those in charge scrambled to find places to put the Aleuts. Ships arrived at each village with no advance notice, and most people were given only enough time to each pack a suitcase.

The 83 Atkans were taken to Nikolski and loaded onto a ship with 477 Aleuts from the Pribilof Islands. In crowded, unsanitary conditions they sailed across the Gulf of Alaska and into Southeast, where they disembarked at old canneries at Funter Bay on Admiralty Island and nearby Killisnoo Island.

The remaining Aleutian islanders – from villages of Nikolski on Umnak Island; Kashega, Makushin, Unalaska and Biorka on Unalaska Island; and Akutan on Akutan Island – were also shipped to Southeast. The 111 Unalaska villagers ended up at another abandoned cannery on Burnett Inlet on Etolin Island, while the remaining 160 Aleuts were taken to an old Civilian Conservation Corps camp at Wards Cove near Ketchikan.

Conditions were deplorable at each site, with inadequate housing, sanitation and medical facilities. The Aleuts had few supplies or provisions and lacked hunting and fishing gear, boats and other important cultural tools.

Meanwhile back in the Aleutians, U.S. soldiers prepared to retake Kiska and Attu. Special forces of Alaska Natives, including Aleut soldiers familiar with the terrain and conditions, scouted the western islands in advanced landings. The Americans built an airfield

A collapsed World War II building marks this view of Nazan Bay on Atka. (Harry M. Walker)

on Adak to increase bombing of Kiska and Attu and built a fighter airstrip on Amchitka.

On May 11, 1943, U.S. troops secured a beachhead on Attu. They fought the Japanese for 18 days in blood baths that made Attu one of the most costly assaults in the Pacific. Fighting ended with a suicidal attack by the Japanese. Of the 15,000 U.S. soldiers on Attu, 549 died, 1,148 were wounded and about 2,100 suffered from sickness and non-battle injuries. The Japanese lost their entire garrison: 2,350 died and the remaining 29 were taken prisoner.

U.S. forces then landed on Kiska to find it empty. Some 5,100 Japanese soldiers had already escaped that island in a shroud of fog.

For all practical purposes, the Aleutian campaign was over. The island bases could still support attacks on Japan's Kurile Islands, but the direct threat was gone with the enemy's retreat. Yet another year passed before the Aleut villagers were allowed to return to their island homes.

During their three-year internment in Southeast, one of every 10 Aleuts died, mostly from disease in the camps. The survivors returned to find little left. Their villages and homes had been occupied and trashed by soldiers and had fallen to ruin through neglect in relentless pounding by weather. The Aleuts received about $35 each from the government for loss of personal property, and rebuilt their villages and homes with additional government money and supplies.

Few people outside the Aleuts knew anything about this until the late 1980s, when Congress took up the issue of compensating Japanese-Americans, also interned during the war. The Aleut evacuation became part of these hearings. In 1988, Congress finally authorized $27 million in reparation payments to the Aleuts. Each of the 400 living survivors of the evacuation received some $12,000, about $16 for each day in the internment camps.

Biologist Edgar Bailey inspects the hulk of a Japanese mini-submarine abandoned since World War II on Kiska Island. (Edward Steele)

A trust fund since established with $5 million of the reparation pays out $500 a year in nontaxable distributions to Aleut senior-citizen survivors of the camps. Work rebuilding six Aleutian churches was scheduled to start summer 1995. The trust also donated $25,000 to an Aleut scholarship program administered by the Aleut Foundation, a nonprofit arm of the Aleut Corp., one of the 13 regional Native corporations established by the Alaska Native Claims Settlement Act of 1971.

Borge Larson

By Charlie Ess

EDITOR'S NOTE: *A free-lance writer and correspondent for* Pacific Fisherman, *Charlie has known Borge Larson for many years.*

Borge Larson looks out his dining room window to the new community center named after him. To its left, a narrow gravel road leads up the valley, past the False Pass clinic, the school, to a housing project, all tokens of a leader and his calculated conviction.

That he would become mayor once the Aleutians East Borough accepted False Pass as a second-class city in 1991 was in some ways inevitable. His round face brightens when he shrugs off the fact that he's faced no opposition and has been reappointed each year by the other six members of the village council. Then again, it seems only fair that someone willing to come all the way from Denmark to carve out a life in an Aleut settlement of a dozen should bask in a little notoriety when the community grows to a city of 90, complete with vehicle

Born in a country whose people live close to the sea, Borge Larson came to the United States in 1949, to a hog ranch in the Midwest. Heartland ranching wasn't for this seafarer, so he followed his uncle north to the Aleutians, where he has been a fixture in the False Pass landscape for more than four decades. (Courtesy of Charlie Ess)

registration, a public safety officer and a new city dock.

But political stature is easy to forget in a man who for the past 45 years has been many things to many people. He has been Santa Claus to Aleut children at Christmas. He has been an old friend at the end of the dock to mariners stopping on their way through Isanotski Strait, where the waters of the Pacific Ocean first mix with those of the Bering Sea. He has been the town's notary public and a winter watchman at the Peter Pan cannery. But as far back as most people can remember, he and his black Lab have been fixtures in the town, at least for as long as the town has had weathered boardwalks.

He and his wife, Lolly, live in a white house with blue trim. A picket fence attached to the front of the house keeps loose dogs from digging up a miniature potato patch. A satellite dish graces his yard. Out back, ducks preen themselves near a homemade watering hole. Chickens scratch at the gravel under an open window where a freshening southeast breeze parts the curtains.

We're having coffee, as much a nightly ritual with the Larsons as driving their Bronco to the end of the airstrip to watch a whale breach or mallards nest, or a bear amble along the beach. Lolly has baked fresh cookies and we're looking back in history. Borge has captured much of it in black-and-whites and color slides. An evening spent perusing photo albums at his dining room table becomes a late-night excursion in the town's chronicles. Pressed among pages of whales, bears and volcanoes are pictures of children long before they grew up to acquire houses, spouses and children of their own.

He sweeps his stout fingers over the worn pages all the while adding commentary in a deep Danish accent that hasn't much diminished in his years among Aleut fishermen. He

Aleutian Island residents have always looked to the sea for their food, and in prehistoric times for their clothing as well. Marine life still supports the bulk of the region's economy. (Courtesy of Charlie Ess)

too hasn't escaped documentation. With sky-blue eyes, magnified slightly behind his glasses, he gazes back to the era when he didn't need them, when his head of silver hair was blond, to the years before cataracts, when he craved a good cigar and a drink instead of a daily dose of insulin.

Born in 1921 in Denmark, Larson showed a yearning for destinations elsewhere when at 16 he found work aboard a three-masted sailing schooner bound for Spain with 540 tons of salted cod. "The trip took 56 days," Larson says as if it were yesterday. He left Denmark for the United States in 1949. World War II had ended, along with his time in the Danish Navy that included eight months as a German prisoner of war in Copenhagen. "I came over here to be on the water, to fish and do things like that."

Traveling inland to the Midwest, he found work at a hog farm in Iowa. "That wasn't my bag of beans," he says, his laughter rising to a hearty cackle. "No salt water," was his biggest problem with the place. "I had been around salt water all my life." His complaint ended a year later with a stateside visit from his uncle, Kris, who'd come to Alaska in 1916 and had become a cannery foreman at False Pass. "I hit him up for a job," Larson says. The following spring he boarded a steamship that took him from Seattle to Seward, then rode on the *Garland*, a mailboat that stopped at False Pass.

Larson found comfort at the sight of Unimak Island. The country reminded him much of his homeland, a place he wouldn't get back to for another 17 years. He got his first real taste of its differences, though, when an earthquake jolted

him awake a few nights later. "The earth was rattling and shaking. I looked and this mountain up behind here was all afire. I thought, 'Boy, is that blowing up now.' It turned out it was rocks rolling down the hill, making sparks. That was a nasty one."

That summer he worked as a trap watchman on P.E. Harris No. 8, one of several salmon traps south of the village in Ikatan Bay. "I was supposed to work on the beach gang, but they found out I could mend web and stuff like that, so they put me on the web gang on the traps."

The U.S. Fish and Wildlife Service regulated commercial fisheries in the years before Alaska won its statehood. "We started the 27th of May

and it (the season) closed the 3rd of August." Those dates were rigid as was a weekly fishing schedule. As a means of enforcement, trap watchmen were issued a season's worth of numbered locking seals. At 6 p.m. each Friday, Larson was to shut the entrance to the funnel-shaped web tunnel where fish entered the trap and affix a locking seal that then had to be cut on Monday morning to open the tunnel and resume fishing. An exception came one Sunday, however, when a changing tide formed up a rip teeming with fish. Soon, thousands of sockeye salmon enticed Larson as they collected in the gin-clear water and swam along the chicken wire lead extending 800 feet from

shore. "The start of the flood was when the trap loaded up." He earned "day money" plus $12 for every 1,000 cases of canned salmon that No. 8 produced.

Then Larson discovered that even with the seal in place the tunnel could be partially opened. That discovery resulted in 57,000 fish being caught in the trap in two hours and kept two tenders busy brailing a good part of the next day. Fishing the traps was a good life, Larson says. But in 1957, two years before statehood did away with the docklike structures, he became beach boss at the False Pass cannery, which had a growing fleet of seine boats.

It was then he began a 22-year stint as a postal clerk and joined a newly forming village council. He took particular interest in establishing a school, which many believe lured families into False Pass from the outlying settlements of Ikatan, Sourdough Flats and Sanak Island. "The school is everything. When there's no school, there's no town," Larson says. "And I know a lot of people in town here feel the same way." When the village's first school, a one-room building, needed funding for electrical power — and neither the fish packing company nor the state would contribute — Larson took it upon himself. "To hell with you," was his reaction. "I'll buy a seven-and-a-half kilowatt Lister (generator)." And that he did, providing for the school's electricity the following nine years.

But school-related issues haven't been as tough on the town as its funerals, Larson confides. Services have remembered loved ones from nearly every family, including his own with the loss of his first wife, Evelyn, in 1973. This becomes an obviously painful topic, an uncomfortable line of questioning. His voice grows softer; his eyes dart from the ceiling to the table and around the kitchen, and he talks with intermittent pauses, a reverence of sorts, while recounting friends lost to guns, the sea and a bear. Too often he has been the town's bearer of bad news.

Offsetting such memories are the pranks, like the time an Italian newcomer joined the cannery crew and stayed on through winter. Hearing tall stories of halibut caught off the end of the cannery dock in summertime, he faithfully set line each day. "There's no fish in winter," Larson says, laughing. "But he didn't know that. So, one day I tied a frying pan on the end of his line. Pretty soon I hear him yelling, 'I think I got something.'"

"Yeah, by golly, you got the pan," Larson says he told him. "Now all you need is the fish."

The punch of many jokes those years rode on ethnicity. Norwegians worked on some traps, Swedes worked on others. And in the midst of it all was Borge the Dane. Larson relishes one of the area's earliest marine radio conversations between two Norwegian skippers, one of whom was manning the *Trojan* when it ran aground with a load of salmon near False Pass. Larson easily recalls the desperate skipper hailing the *Kathy B*, a nearby tender, whose call sign was WWJ: "Calling wubble U wubble U yay."

"This is wubble U wubble U yay," came the response from the *Kathy B*.

"Wubble U wubble U yay? This is the *Troyan*. This is the *Troyan*. This is uryent." According to some, Larson himself grappled with the English language when he arrived, and for a time mumbled only "Hullo" to passersby on the docks. But that was a 39-year cannery career and nearly three black Labs ago.

Tonight, he answers his cordless phone several times during the evening. In the living room his TV is tuned to The Discovery Channel. Duke Jr. lies outstretched on his side nearby; his thick tail whips the floor in acknowledgment that we've made eye contact.

Two of the 26 kids enrolled in the school stop by to get the keys to the school gymnasium.

While being the mayor could be dismissed as a plaid-shirt position and concerns many days lie with damage caused by the region's storms or nuisance bears prowling the beach, other issues are more pressing, like finding investors for the new, 400-foot dock. Because False Pass is a stop for the state ferry serving southwestern Alaska and because the port has accommodations for larger vessels, the town is exploring options of tourism and freight storage to augment its commercial fishing economy.

When his days aren't taken up with small-town politics, there's always the threat of plate tectonics. When Westdahl Peak erupted in 1991 – and townspeople feared that the 1,600-square-mile island would split in half and slip into the sea – Larson and a few others char-

The salmon purse seiner Perseverance, *owned by local fisherman Melvin Smith, was a regular sight around False Pass until it burned in late summer 1994. (Gilda Shellikoff)*

tered a small plane from Cold Bay, a 20-minute flight away on the Alaska Peninsula mainland, then flew over the fissure, cameras clicking, to record molten lava and a plume of ash.

Wednesday nights are activity nights in the Larson Center. "There's cribbage, coffee and cookies," Larson says. "There's a pool table, a Ping-Pong table and toys for the little ones." In the other end of the building are the offices where he and the council conduct the city's business and plot its future. It is there that Larson can "make a plan, follow up on it and keep on pushing."

The Aleutians Today

By L.J. Campbell

Economy

For all that has happened, the Aleutians remain one of the least charted and most sparsely populated regions of Alaska. About 5,017 people live here year-round in five towns in the eastern and central islands, with another few hundred occupying military bases in the western islands.

As it has always been in this part of the world, what comes from the sea shapes what happens on shore. Subsistence fishing and hunting continue to be important, particularly in the small villages where jobs are few and groceries, delivered by barge or plane, are expensive. Subsistence is also one of the strongest links to Aleut cultural traditions.

Commercial fishing in the North Pacific and Bering Sea drives the Aleutian cash economy. In most of Alaska, commercial fishing means salmon. Out here, it means crab and bottomfish such as pollock, cod and halibut.

Commercial fishing has been going on around the Aleutians for decades, but recent years have brought big changes. In 1980, the U.S. government claimed ownership of the water within 200 miles of American shores, and banished the foreign fishing fleets that had formerly dominated. Now, large American vessels do all the fishing and crabbing, delivering their catch to floating and onshore processors.

An outcome of this has been Unalaska's transformation from quiet Aleut village into boomtown. When foreign seafood companies could no longer work in American waters, they headed to shore. Several of them landed in Unalaska where they invested billions

The commercial center of the Aleutians is the port at Dutch Harbor on Amaknak Island in Unalaska Bay. Commercial fishing drives this commerce, with Unisea Processing at lower left in this image and Alyeska Processing at center right. In the middle is Standard Oil Hill. (Barbara Keller)

Air transportation is another crucial link in the development of an expanded economy. During the fishing season the Dutch Harbor airport is a busy place, but transportation to smaller Aleutian communities is much more problematical. Weather, cost and frequency of flights are key elements in marketing the smaller communities to tourists. (Harry M. Walker)

of dollars in new processing plants. Unalaska's Dutch Harbor is the busiest commercial fishing port in the nation and one of the world's major fishing centers.

Meanwhile, more and more boats and processors have been coming into the fishery. The government continues imposing controversial fishing regulations to allocate the stocks among the many users. With the increased fishing effort, seasons continue growing shorter. This creates uncertainty about the industry's future and makes many in Unalaska uneasy. Efforts to diversify the town's economy include promoting it as a tourism gateway to the Aleutians.

In another fishing spinoff, the government allocated a small percentage of the Bering Sea cod harvest to Native villages along the coast. Otherwise, local fishermen were being excluded from the wealth at their front doors because their boats were too small to fish the high seas. The villages set up community development associations and formed partnerships with seafood processors to harvest their quotas. The Aleutian communities are part of the Aleutian Pribilof Island Community Development Association (APICDA).

Through APICDA, Aleut villagers are finally seeing some payoff from bottomfishing. Trident Seafood Corp., an American-owned processor with a plant in Akutan, harvests and processes APICDA's cod, and finds jobs on fishing trawlers or processors for APICDA villagers. With the profit from cod, APICDA has built village docks and cold storage facilities, and bought larger boats for local fishermen.

Further, Native fishing associations successfully lobbied for a special halibut fishery, using jig gear (hooks on a line operated by a diesel winch) from their small boats. Their catch brings a premium price when they can load it on planes for the fresh-fish market in Anchorage.

Clearly, many of today's Aleuts navigate the tricky swells of politics and business, as well as the sea.

Corporate culture hit here after Congress passed the Alaska Native Claims Settlement Act in 1971. Aleuts suddenly became shareholders in the Aleut Corp. Today, the corporation owns real estate in Anchorage, an oil drilling company in Cook Inlet, a construction subsidiary and a Colorado-based service-contracting subsidiary. It also owns about 70,000 acres of subsurface rights in the chain, which include some minerals but consist mostly of areas of geothermal activity from volcanoes. None of that has been developed.

Unisea opened the Grand Aleutian Hotel in 1993 on Margaret Bay, a tiny offshoot of Iliuliuk Harbor. With uncertainties in the fishing industry, Aleutian Islanders are seeking to broaden their industrial base and look to tourism as one aspect of that effort. With the increased accommodations, organized land and sea tours have been developed for visitors that focus on bird- and marine-mammal watching, fly-in sport fishing, history and commercial fishing and fish processing. (Alissa Crandall)

The Military's Legacy

The U.S. military dug deeper into the Aleutians after World War II, as allied relations with the Soviet Union dissolved. The western islands became the domain of spy planes, submarine chasers and secret intelligence listening posts as the United States took on Cold War enemies.

A Navy station on Adak and an Air Force base on Shemya supported U.S. air and sea patrols and intelligence gathering into the early 1990s. A long range navigation aid (LORAN) station on Attu transmitted locator signals.

The Aleutians also attracted weapons developers

seeking remote test sites. Three underground nuclear explosions, including a five-megaton warhead, the most powerful nuclear device ever detonated on American soil, shook Amchitka Island in the late 1960s and early 1970s.

Although Amchitka was uninhabited and 200 miles from the nearest community, the testing triggered protests. People worried about tidal waves in the Pacific and release of radioactivity. The third shot, Cannikin, fired nearly 6,000 feet below the surface, measured 6.8 on the Richter scale. Rock cliffs crumbled and the ground collapsed into a 40-acre lake above the blast. About 1,000 sea otters died from

A sign warns of polluted water, a legacy of World War II on Adak. (Lon E. Lauber)

shock waves in the water, but no tidal waves occurred.

No significant amounts of radiation have ever been detected in surface monitoring, although traces of tritium, a radioactive form of hydrogen, are found in water and soil above the first blast site. These traces are within acceptable federal levels. Questions remain about radioactivity deep within the island and what, if any, amounts of radioactivity are leaching into the ground water and moving out into sea.

The U.S. Fish and Wildlife Service in 1995 was coordinating cleanup from military activity on national wildlife refuge lands throughout the Aleutians, including Amchitka. The federal energy department may do more extensive testing at Amchitka while the Environmental Protection Agency was deciding whether to put the island on its Superfund list of hazardous sites eligible for federal cleanup money.

The naval installation on Adak already is listed on the Superfund. Cleanup is also underway at Shemya, although that base is not a Superfund site.

During the war, these bases were hastily constructed as forward staging areas for raids on Japanese-held Kiska and Attu. The Navy retained the Adak base, while the Air Force used tiny Shemya Island as a refueling stop during the Korean War, then later as a base for strategic intelligence-gathering. During the 1970s and 1980s, both bases grew into important overseas security commands.

At Shemya, where the military occupies the entire island, the Cobra Dane radar facility was built in the mid-1970s to monitor space and missile activities, collect intelligence information and track earth satellites. The base was named Eareckson Air Force Station in 1993, and at times it has held up to 900 personnel.

On Adak, the Naval Air Station provided full support for West Coast patrol squadrons, including P-3 Orion planes doing anti-submarine patrols and surveillance flights, and C-130 cargo and transports.

A Navy security group that monitored military transmissions throughout the Pacific also operated on Adak.

As the Navy and Air Force poured millions of dollars into the Aleutians, they upgraded these remote bases into modern urban centers. Adak celebrated when McDonald's opened in 1985, in a ribbon-cutting of 101 dollar bills. A $23 million middle-high school opened in 1992, and the next year its boys basketball team won the state AA championship. Caribou transplanted to the south end of the island provided sport hunting and Adak personnel could go halibut fishing on military boat charters.

Shemya gained a new gym in the late 1980s. Satellite TV brought California programming, first-run movies came on cargo planes, weekly mail planes brought letters from home, and military telephone lines gave each person a 20-minute daily "morale call" to bases Outside, where they could be routed onto local lines for low-cost conversations. Such things were important in a place where wind conditions could keep everyone inside for hours, perhaps days. A new fire truck being unloaded from a ship was blown off an icy dock in one such gale, and was never recovered.

The dissolution of the Soviet Union essentially ended the Cold War, and the U.S. military started downsizing. In 1993, the Navy quietly dismantled an $86 million over-the-horizon radar site built four years earlier on Amchitka. In 1994, it downgraded Adak Naval Station by closing services, moving out families and reducing staff to about 500 people. In 1995, Congress was considering closing the naval facility completely. Also at Adak, the Naval Security Group Activity communications station is trimming operations in September 1995 with complete closure by 1996.

The Air Force station on Shemya was also shutting down in 1995. A civilian contractor will maintain the facilities and keep the runways open for emergency use.

Agattu Island rises in the distance in this view from atop the Cobra Dane radar facility on Shemya Island. (Cary Anderson)

In the meantime, the U.S. Coast Guard continues staffing the LORAN station on Attu. A plane comes in every two weeks with supplies and mail for the crew of 15.

Winter winds have been clocked at 150 knots at this westerly tip of the Aleutians; in late winter 1995, winds blew the station's wind meter off its post. In more peaceful interludes, people go skiing and hiking, enjoying the scenery and wartime historical sites. Spring brings flocks of migrating birds and groups of birders. The island is still littered with unexploded ordinances from the war, so until cleanup is complete, movement is restricted to the national historical monument about five miles in radius around the LORAN station.

FACING PAGE: *A world war and subsequent military presence has earned the Aleutians their greatest notoriety in the 20th century. As the century comes to a close, however, the Adak Naval facility is scheduled to shut down and Air Force personnel will leave Shemya. The Coast Guard LORAN station on Attu faces an unknown future. The Attu station has about 20 people and a pet St. Bernard dog named Coco that has lived here about 10 years. The guardsmen stay only a year, with a 30-day leave. Coco gets off the island occasionally to visit veterinarians in Kodiak or Anchorage. Here a Coast Guard cutter pulls into Sweeper Cove on Adak. (Lon E. Lauber)*

RIGHT: *Many residents of the Aleutians are striving to resurrect traditional Aleut culture and its modern interpretations. Daniel Shellikoff, from False Pass and a student at the University of Alaska Fairbanks, shows off an Aleut mask he made. (Gilda Shellikoff)*

The Communities
FALSE PASS

Unimak, largest island in the Aleutians, also used to be one of the most populated. Russians reported 12 Aleut villages in 1840, and the island had several settlements as late as the 1940s. Today the island's only community is False Pass, a town of about 90, located on the eastern shore.

The town faces Isanotski Strait. Its Aleut name "Isanax," meaning "the pass," became False Pass when early mariners thought the channel impassable. Today the strait, or "bay," is a well-traveled marine highway.

Most everyone in False Pass is Aleut. Fishing is a way of life here. The kids grow up doing it. At their modern school, where basic education is supplemented by computers, one of their projects is writing about and photographing fish to put on CD-ROM. They get Spanish, physics and other enrichment classes on TV, beamed to them by satellite in a federally funded program. They play basketball, volleyball and do cross-country running. But they all help out when fishing season rolls around.

Most families fish commercially in summer for salmon, their main cash crop. False Pass is near the heart of one of the state's biggest salmon fisheries, the South Unimak district of Area M. This is often incorrectly called the False Pass fishery, because its fish used to be delivered to False Pass.

Outside salmon season, fishermen haul in halibut, cod, crab and sometimes octopus. Some of this is sold, but it's also dried, boiled, baked, pickled and eaten. One person even hunts fur-bearing sea otters, now plentiful in the bay.

Homesteader William Gardner settled here in the early 1900s, but it wasn't until 1917, when gold-mining magnate P.E. Harris opened a salmon cannery, that a town started forming. The cannery drew workers from area villages, such as Ikatan on the island's Ikatan Peninsula; Morzhovoi on the Alaska Peninsula; and Pauloff Harbor on Sanak Island. The False Pass post office opened in 1921. In 1951, Yugoslavian fisherman

This view of False Pass shows the Peter Pan docks and other buildings of the tiny community. The blue-roofed building is the school. The photo was taken from across what residents call "the bay," which refers to Bechevin Bay off the Bering Sea. Bechevin is connected to Ikatan Bay on the Pacific side by narrow Isanotski Strait. (Gilda Shellikoff)

Nick Bez bought the cannery, the first in a string of acquisitions that became Peter Pan Seafoods. Bristol Bay Native Corp. bought the company in 1972, then sold it in 1979 to Nichiro Gyogyo Kaisha of Japan, its current owner.

In 1981, the cannery burned, taking about 130 jobs.

Although the processing facility wasn't rebuilt, Peter Pan continued running its camp with food and gear storage for fishermen. Its grocery store, docks and fuel depot serve as a seaside quick stop for vessels.

Today False Pass is reaching beyond its former identity as a company town. The village corporation, tribal council and city are trying to create business opportunities so local people can have jobs and the young people will stay. In 1992, the city completed a $1.8 million dock, funded by the Aleutians East Borough, state and federal agencies, and local tax money. It's designed for fishing and crab boats, floating processors and Alaska state ferries. The city wants to develop an industrial marine park on adjacent land.

False Pass is also part of the Aleutian Pribilof Island Community Development Association, which harvests a share of Bering Sea pollock. With $400,000 from APICDA, the city extended water and sewer to its new dock. APICDA also provided $200,000 to the tribal council to build a gear storage warehouse.

The recent construction inspired the children of False Pass to build a toy boat harbor in a creek near town. They float homemade cork and wood boats, mooring them at miniature docks with warehouses made of wood scraps. When storms come, in the true tradition of fishermen, they go to their harbor to make sure their boats are safe.

AKUTAN

People in Akutan tune to channel 6, the local chat line on their VHF radios. There's a lot to talk about. Construction projects with local jobs. A new hydroelectric plant. A $1.6 million remodel of the school. Eight new homes going up and 16 older ones getting electric water heaters and new roofs.

Some roofing blew off in the last storm, someone reports. The shingle manufacturer seemed surprised; the shingles were guaranteed to withstand 50 mph winds. Chuckle. Akutan storms blow more like 80 mph. Later word comes, a skiff has landed with fresh sea lion. Get to the beach if you want some.

Town leaders are working on more for the community to talk about. They want a boat harbor. It's Akutan's most pressing need, they say.

Akutan, a town of about 90 year-round residents, sits beside a mountain on the shore of Akutan Harbor, a 4-mile-long fiord at the island's east end. It's an old Aleut

Youngsters in False Pass spawned their own fleet in a creek near town. The boats are made of cork and packing material, the dock of found wood. (Gilda Shellikoff)

community, founded about 1879 by people from surrounding islands, helped by Russian Orthodox priests.

Akutan Harbor made a natural port, located near Unimak Pass, a busy shortcut between the Pacific Ocean and Bering Sea. A fur trading post moved in, and its agent started a cod fishing business.

In 1912, a Norwegian whaling station opened at the head of Akutan Harbor. Most Akutan men worked there, rendering whale oil and cooking meat into fertilizer. The work on oily planking around sharp butcher knives and hot vats was dangerous. Aleck McGlashan, now 93, of Anacortes, Wash., worked in the plant as a teenager. His family moved to Akutan in 1907, when his dad, Hugh McGlashan, bought the store. His father, from England, and mother, from Attu, met at the cod fishing station on Sanak Island. They had 13 children. The villagers lived in barabaras

Akutan village sits on the shore of Akutan Harbor, about a 20-minute floatplane ride from Unalaska. Houses and businesses in Akutan are connected by boardwalks instead of roads, so most people walk to where they need to go. (Alissa Crandall)

then, Aleck said. People fished in summer and trapped foxes in winter.

Aleck became a gunner on the whaling boats. The 150-foot steam-powered vessels chased humpback, fin, sperm, blue and right whales. He earned $60 a month, 50 cents an hour overtime, plus bonuses of $2 to $10 for each whale. The plant closed in 1942, during World War II.

After the war, fishing revived. The late 1970s crab boom brought seafood processors to the bay.

In 1980, Trident Seafoods opened a shore-based plant a quarter-mile west of town. Today it's one of the biggest volume processors in Alaska, handling cod, pollock, crab and halibut delivered by a fleet of boats. It employs about 800, mostly Filipino, during peak season.

The people of Akutan are trying to become more involved with the industry at their front door. The city completed a new dock in 1991, with a gear warehouse and a supply store for boats. The dock will accommodate the state ferry and perhaps someday tour boats. But it's not equipped for offloading fish.

Neither is it protected from wind-driven storm waves. And because they don't have sheltered anchorage, Akutan's fishermen use 16- to 24-foot skiffs that they can pull onto land in winter. If they had a boat harbor, they could own bigger vessels, more suited for deep-sea fishing.

The city conducted a survey and found that more than 300 boats, mostly out of Seattle, delivering to processors in the bay would use permanent and transient moorage in Akutan if available. The city is trying to get funding from federal and state agencies to build a $6.1 million boat harbor with a permeable wave barrier. Akutan's only local source of income is a 1 percent fish tax.

So Akutan fishermen dabble in the bottomfishery, longlining for halibut and cod from their skiffs outside the bay. Sometimes they crew on crab boats and occasionally are hired at the processor when openings occur. In summer, some of them crew on salmon and herring boats in Alaska Peninsula and Bristol Bay fisheries.

Getting in and out of Akutan usually involves a floatplane; the town has no landing strip. Akutan fishermen sometimes drive their skiffs to Unalaska. During the school year, Akutan students fly to other towns to play basketball. They go to Dutch Harbor for a week of intensive swimming lessons and enjoy the larger town, including fast food at Burger King.

Akutan has its store, a cafe, a roadhouse, a bingo hall, a day care center, a library. The city plans to build a multi-purpose building with a new, larger library, recreation center and museum.

The Akutan Traditional Council is trying to keep Aleut culture alive. It offers wood carving classes that include stories about the items' traditional uses, holds summer spirit camp for youngsters and plans to stage plays with Aleut themes.

With a state grant in 1993 the council published *Hunters of the Sea Akutan Cookbook*, dedicated to Luke Shelikoff, Akutan's last traditional chief who died in 1982. The cookbook speaks to the community's seaside heritage, with recipes such as Seagull Egg Pie, Halibut with Minced Crab Tails, Octopus Patties, Salted Seal Soup and Swiss Sea Lion Steaks.

People wait for the state ferry to arrive at Akutan's new dock, completed in 1991. The community now is trying to get a boat harbor with a breakwater, so their vessels can be sheltered from wind-driven waves. (Alissa Crandall)

UNALASKA

Hiking through wildflowers on the mountains above Unalaska, particularly during a drizzle, brings an incomparable feeling of calm. Return to town, though, and hold on for a ride.

A whirlwind of activity engulfs the City of Unalaska and its Port of Dutch Harbor. With more than 4,300 residents, it's the biggest town in the Aleutians. Its industrial seafood processing capabilities are unmatched in Alaska. Its port ranks first in the nation, in volume and dollar value of fish unloaded. City officials estimate that their community serves

In 1912 Norwegians opened a shore-based whaling station at Akutan Harbor. The company took 310 whales its first year of operation, but as time passed the industry basically died out in Alaska, even though the Akutan plant didn't close until 1942. (John Johansen)

upwards of 15,000 people each year, counting seasonal factory workers and visiting fishermen.

With this, Unalaska has grown into the transportation and freight hub of the Aleutians. Jets stop with breathtaking abruptness on the city's short seaside runway, after steeply descending between mountains. Meanwhile, barge traffic grows in the harbor, with bigger and more frequent container ships visiting port.

Unalaska experienced its first seafood boom in the late 1970s, with the explosion in Bering Sea crab fishing. A second boom convulsed the town in the late 1980s when changes in the cod and pollock fisheries drove foreign seafood processors to build new plants on shore. Today, six major processors and several smaller ones churn out millions of pounds of surimi and other seafood products each year. While most heavy construction has slowed, rock blasting and dirt moving continue.

The town is catching up with its industry. A new boat harbor is proposed. The city is considering buying electrical power generated by harnessing Makushin Volcano's geothermal energy. The Makushin project, promoted by a private firm, is still in the planning stages. Downtown, the tiny old clinic has been replaced with a spacious new one. There's a new city hall, a new community center, a new school addition, and a new post office, motel and restaurant. A new pedestrian trail will wind from homes in the valley, along the bay and through town, and across to the port when it's completed. Coming summer 1995...paved streets.

On Amaknak Island, the city's commercial district that connects to downtown by bridge, a toney new development surrounds Margaret Bay. The centerpiece Grand Aleutian Hotel, with its upscale restaurant featuring "North Pacific Rim Cuisine," is working with the new Unalaska Convention and Visitors Bureau to market the island to tourists. They're aiming at

adventurous types who want fly-in fishing trips and birding boat tours. The city and the Ounalaska Corp., the village corporation and the town's major land-owner, are putting together historical and cultural tours.

The Margaret Bay complex also has a new post office, as well as two new supermarkets and two banks. Alaska Commercial Co.'s new building holds a Burger King; a Kentucky Fried Chicken and Pizza Hut are in the works.

Fast times, indeed, have come to Unalaska.

The center for commercial activity in the Aleutians, Dutch Harbor has the busiest port and airport in the region. But in a landscape consisting mainly of mountain tops, as seen in this view from 2,136-foot Pyramid Peak, finding sufficient level ground to lay out a runway has created an engineering challenge for this community 800 air miles from Anchorage. The runway is visible at the foot of Mount Ballyhoo on Amaknak Island. The port follows the shore of the island, which is connected to Unalaska Island by the bridge at left center. The town of Unalaska sits at right center between the bay and the lake. (Scott Darsney)

LEFT: *The recent fishing boom and a desire to broaden their economic base by encouraging tourism has prompted Unalaska residents to add some amenities to their community. The town park, a planned museum, new community center and new lodging and meal facilities enhance traditional fishing and hiking pastimes. (Harry M. Walker)*

LOWER LEFT: *Makushin Volcano towers over this view of downtown Dutch Harbor from the summit of 1,651-foot Mount Newhall. Planners have cited the volcano's geothermal energy as a potential source of power for Unalaska and Dutch Harbor. (Scott Darsney)*

Being the center of activity is not completely new to Unalaska, although there have been times when nothing much at all went on. Some recall the lulls with fondness. People fished, foraged and hunted along the water, without the crowds or pollution that mark the bay today.

Even so, Unalaska was one of the most populated islands, a center of Aleut culture, when the Russians arrived. It had the first permanent Russian settlement in Alaska, called Iliuliuk. Unalaska became the Alaska headquarters for the Russian Orthodox Church.

Under the Americans, Unalaska and its Dutch Harbor port became an important Bering Sea outpost. A customs agent was assigned, trading companies opened stores, ships stopped for supplies, fuel and mail.

World War II brought wholesale changes to Unalaska's landscape and people. The military blanketed Unalaska with acres of buildings. Its complex extended into the valley and up the mountainsides, where bunkers and gun turrets created square peaks. Most of the old military structures are gone today, but miles of roads remain. The war also scarred the Aleuts, who were evacuated to Southeast. Of the island's four prewar villages, only Unalaska was resettled.

NIKOLSKI

The tiny Aleut village of Nikolski sits on a nape of land facing the Bering Sea near the western tip of Umnak Island.

This is an ancient site, one of the oldest continually inhabited places in the western hemisphere. Artifacts recovered at nearby abandoned Chaluka village date back 4,000 years. Older signs of habitation, some 8,000 years old, are found just across Nikolski Bay on Anangula Island.

Nikolski now faces a challenge, to survive and keep its Aleut heritage alive.

In 1994, Nikolski's year-round population had dwindled to 24 people, many elderly. The school had closed for lack of students. The post office also closed, when the postmaster retired after 40 years with no replacement.

The villagers like living here. It's quiet and pretty, though often foggy, and they can fill their freezers without buying a lot of expensive groceries. They fish and hunt ducks, geese and sometimes sea lions from their dories. They occasionally hunt reindeer, put on the other end of Umnak in the early 1900s. They also butcher a few wild cattle and sheep, offspring of stock brought by Russians and later Americans.

Earlier this century Nikolski was busier. A ranching operation ran about 15,000 sheep and several hundred cattle, hiring villagers for round-ups and shearing. The ranch supplied meat and milk during the Depression of the 1930s, when groceries and ammunition for hunting were hard to get, remembers villager Simeon Pletnikoff, 76.

In the 1950s, a Distant Early Warning station brought about 40 jobs and Nikolski grew to about 100 people, said Val Dushkin, 63, the retired postmaster. The station closed and the ranch folded in the early 1970s, when Chaluka Corp., the village corporation created by the Alaska Native Claims Settlement Act, acquired 69,000 acres around Nikolski and bought the livestock.

Today, the corporation operates the village's only store. A Russian Orthodox priest visits several times a year.

Villagers say they want Nikolski revived. That means bringing in young people with children, but there are few jobs to attract them. "The village of Nikolski is fighting for its life," says a report by APICDA.

A 1994 tourism study assessed Nikolski's potential as a tourist destination. The biggest obstacle to developing tourism is access. Weather frequently delays flights from Dutch Harbor and the shallow bay prevents large boats from coming close to shore. If a suitable vessel could be found, the town has some accommodations and plenty of natural attractions to lure visitors. (Aleutian Housing Authority)

FACING PAGE: *The Church of St. Nicholas is the latest in a line of Russian Orthodox churches reaching back to 1806 when the chapel at Nikolski was located in a barabara at the site of the prehistoric village on Chaluka mound. Even then the religion was entrenched in the village because the village chief's nephew was baptized in 1759, making him the first Aleut to become a Christian. (Kaarina Stamm)*

RIGHT: *The Aleutian community with the most precarious future is Nikolski, population 24, near the southwestern end of Umnak Island. Most of the town's residents are elderly, and there are few jobs to bring cash into the community. One of the few jobs is operating the village store. (Kaarina Stamm)*

"Without economic development, its future is bleak...."

A few things are happening. A Canadian ranching company has bought the village corporation's livestock, 1,100 cattle, 500 sheep and 50 horses, and is leasing village grazing land. Pat Harvie, an owner of Bering Pacific Ranches out of Alberta, said the Nikolski effort is part of a larger operation, with 3,000 head of livestock on Umnak and Unalaska islands and a planned slaughterhouse on Umnak's east end, at the old Fort Glenn air base.

Meanwhile, APICDA is helping Nikolski villagers finance buying into the commercial halibut fishery. APICDA also leases one of its larger boats to Nikolski for cod fishing. The boat operates out of Unalaska, since the village has no dock and Nikolski Bay is too rocky, shallow and exposed for boat anchorage.

Tourism also may hold potential for Nikolski. Agrafina and Scott Kerr, Nikolski's youngest couple, have formed a tour company, Aleutian Island Adventure, catering to people interested in the area's history, scenery and wildlife.

Agrafina's mother, Annabelle Dushkin, works as a janitor at the village clinic and her dad, Willie, is a mechanic at the power plant. Agrafina went away to high school in Kodiak and came home in 1981, a year after graduation. "I wanted to come back and try to help my community," she said.

The Kerrs live in a comfortable three-bedroom house with two dogs and a rack outside for drying fish. Like the other villagers, they put up most of their food. Working for money means leaving the village, sometimes for several months. Scott, an upstate New York native who sailed into Nikolski on vacation one summer, does longshoring and commercial fishing out of Dutch Harbor. Agrafina works for Aleutians West Coastal Resources Area, counting sea lions at nearby rookeries and conducting water-quality interviews with Dutch Harbor seafood processors.

They don't want to see the village dry up. "I'd like to see more tourism. I'd like to get our school back," says Agrafina. "We could have fishing. If we could have more people out here working, the younger generation, I think, would come back.

"I plan to live here the rest of my life. I guess this is where I'm going to pass away."

ATKA

Atka Island punctuates the central Aleutians like a semicolon. Its circular, mountainous north end connects to its curving tail by a narrow bridge of low land. Atka village sits on this bridge, facing Nazan Bay

Located 1100 air miles from Anchorage and 90 air miles east of Adak, Atka is the most isolated of the Aleutian's Aleut communities, and like many Aleutian communities, it operates on Hawaii-Aleutian time, one hour earlier than the rest of the state. (Lon E. Lauber)

off the Pacific side. It's the island's sole community with about 96 people, all Aleut except three schoolteachers.

Atka Island's main village used to be on a peninsula in Korovin Bay on the Bering Sea. Korovinski village became Russian headquarters for the central and western Aleutians and Commander Islands.

Later, villagers moved to Nazan Bay. During World War II, the Navy temporarily occupied the bay to refuel and rearm planes bombing Japanese positions. One evening the Navy sent the Atkans out to their fish camps. The children covered the tents with grass as camouflage against the Japanese planes seen overhead. Later that night, the soldiers torched the village, a wartime tactic to erase landmarks useful to the enemy.

Today, Atka is the most isolated and traditional of the Aleut villages. People depend heavily on subsistence hunting and fishing. They practice Aleut arts such as fine basketry, carving and making bentwood hunting visors like those worn by their ancestors in skin bidarkas. Aleut is still the primary language, spoken by everyone except children, who are learning it in school as a second language, and the three school-teachers.

At the same time, it is also a progressive village, fluent in society's newest technologies. Its school – a district unto itself with about 28 students – is the most affluent in the state, spending more than $55,000 per child in 1994. Almost every student has a computer and they navigate the worldwide Internet to search data bases and communicate by electronic mail.

The bilingual teacher, Atkan Dennis Golodoff, is developing a computer program that combines digital

Young Aleutian Islanders, such as Jennifer Kost, look to take their place in the 21st century, but also hope to carry on the millenia-old traditions of their Aleut forefathers. (Harry M. Walker)

Jimmy Prokopeuff works on a computer at the Atka school. The local school district, consisting of only the Atka school since the school at Nikolski closed, has the highest per student expenditure of any district in the state. (Harry M. Walker)

other Aleut villages, with chronic unemployment.

The village has about 15 commercial fishermen who longline from small skiffs for halibut and cod. Atka also has a month-long commercial salmon fishery, but the runs are sporadic.

As in other Aleut villages, Atka fishermen need larger boats so they can safely venture farther out onto the fishing grounds and catch enough to make a decent living. Swift currents and harsh weather whip the island, making the skiff fishery hazardous.

Atka fishermen face an additional problem of getting their catch to market. No floating processors or tenders venture this far west. To solve this problem, the fishermen formed a cooperative, the Atka Fishermen's Association, several years ago. They built a small seafood processing facility with cold storage, where the fish could be held for market. The good idea turned sour in 1992 when the broker, hired from outside the village to market the fish, absconded with the money. The cooperative operated the next year with a loan from APICDA. The two organizations have since formed a joint venture, Atka Pride Seafoods, and are expanding and upgrading the plant to operate year-round.

Atka built a floating dock, funded by APICDA, so local boats could unload their catch. Atka fishermen are leased APICDA's bigger 32-foot vessels and increased their halibut catch six-fold in one year.

Plans are underway to build a permanent dock, and the community is trying to obtain surplus equipment from the Navy's Adak facility, 90 miles to the west; it's unclear what may happen to these plans when Adak's facilities close.

While fishing is the backbone of the economy, tourism may someday figure in as well, with visitors drawn to Atka's black sand beaches, hot springs, tundra-covered ridges good for hiking, numerous bays for kayaking and its history.

pictures with recordings of elders speaking Aleut. It's not uncommon to see Golodoff headed home on his four-wheeler with his PowerBook, a notebook-size portable computer, tucked under his arm. He is building on linguistic work started years ago by Atkan scholar Moses Dirks.

Atkans also are building a fishing economy, despite their remoteness from the markets. Fishing has always been important here and looks to be a mainstay in the future, a way to create more local jobs in a place, like

Bibliography

Bergsland, Knut and Moses L. Dirks, editors. *Unangam Ungiikangin Kayux Tunusangin, Unangam Uniikangis Ama Tunuzangis, Aleut Tales and Narratives.* Fairbanks: Alaska Native Language Center, University of Alaska, 1990.

Byrd, G. Vernon, George J. Divoky and Edgar P. Bailey. "Changes in Marine Bird and Mammal Populations on an Active Volcano in Alaska." in *The Murrelet*, 61:50-62, Summer 1980.

Cloe, John Haile. *The Aleutian Warriors.* Anchorage: Anchorage Chapter, Air Force Association, 1990.

Fitzhugh, William W. and Aron Crowell. *Crossroads of Continents, Cultures of Siberia and Alaska.* Washington, D.C.: Smithsonian Institution, 1988.

Hersh, Seymour M. *"The Target Is Destroyed."* New York: Random House, 1986.

Hulen, David. "After the Bombs Questions Linger About Amchitka," *Anchorage Daily News.* Feb. 7, 1994.

Kirtland, John C. and David F. Coffin, Jr. *The Relocation and Internment of the Aleuts during World War II.* Anchorage: Aleutian/Pribilof Islands Association, Inc. 1981.

Kizzia, Tom. "Flight From Adak," *Anchorage Daily News.* March 13, 1995.

Morgan, Lael, editor. *The Aleutians.* Vol. 7, No. 3. Anchorage: The Alaska Geographic Society, 1980.

Motyka, Roman J., Shirley A. Liss, Christopher J. Nye and Mary A. Moorman. *Geothermal Resources of the Aleutian Arc.* Professional Report 114. Fairbanks: State of Alaska, Division of Geological & Geophysical Surveys, 1993.

Pierce, Richard A., editor. *Siberia and Northwestern America 1788-1792, The Journal of Carl Heinrich Merck, Naturalist with the Russian Scientific Expedition Led By Captains Joseph Billings and Gavriil Sarychev.* Kingston, Ontario: Limestone Press, 1980.

Proposal for a Community Development Quota. Juneau: Aleutian Pribilof Island Community Development Association, 1993.

Rennick, Penny, editor. *Unalaska/Dutch Harbor.* Vol. 18, No. 4. Anchorage: The Alaska Geographic Society, 1991.

Rizzo, Wayne., editor. *Eagle's Call.* Commem-orative edition. Adak, Alaska: Public Affairs Office of Naval Air Station Adak, June 24, 1994.

Smith, Barbara Sweetland and Patricia J. Petrivelli. *A Sure Foundation, Aleut Churches in World War II.* Anchorage: Aleutian/Pribilof Islands Association, 1994.

Swanson, Henry. *The Unknown Islands.* Cuttlefish VI. Unalaska, Alaska: Unalaska City School District, 1982.

Mailboat Coming!

By J. Pennelope Goforth

Editor's note: *From a long line of seafarers herself, writer/photographer Pennelope Goforth formerly lived at Dutch Harbor and now lives in Juneau, on a boat. She is currently working on a book about John E. Thwaites.*

"HEAVY HEAD WINDS STILL and very little headway," mail clerk and photographer John E. Thwaites wrote in his journal during a storm at sea. "My mail room looks as though the pigs had been turned into it. The entire floor is covered with all kinds of mail. A single letter will travel all over the room by itself." This was Thwaites' first trip on the most westward mail run of the U.S. Railway Post Office (RPO). The mail would be more than two months late this trip in winter 1905-1906 as fierce Aleutian-bred storms battered the mailboat *Dora* across the North Pacific. Then, as now, the weather in the Aleutians determined when that all-important link to Outside, the mail, would arrive.

Consisting of a 6- by 12-foot 'mail closet' near the galley of the small steamship, this particular railway post office encountered neither tracks nor rock slides, although running up on the rocks was a fearsome hazard. Beginning in Valdez, the mail run on the *Dora* headed southwest with regularly scheduled stops in Prince William Sound, Cooks (sic) Inlet, at Kodiak Island, Cold Bay, Chignik, Unga, Sand Point, Belkofsky, the two lighthouses at Unimak Pass, and out into the Bering Sea to Dutch Harbor and

Operated first by the Alaska Commercial Co. as a mailboat and freighter and later by the Alaska Steamship Co., the sturdy, wood-hulled, 112-foot steamship, S.S. Dora, was nicknamed "the bull terrier of Alaska" as much for its reputation of bouncing off every rock between Valdez and Unalaska as it was for its seaworthiness. John E. Thwaites described the Dora, built in 1880 for the Bering Sea fur seal trade, as "very carefully constructed with a view to bucking ice, and riding all sorts of seas in all sorts of weather. [Its] speed fore and after is from 7 to 8 knots, that of [its] side motion has never been definitely determined.... There is not a more slow, staunch or doughty little craft afloat on the waters of the Pacific." (Alaska State Library, John E. Thwaites Col., No. 18-410)

Unalaska, with occasional unscheduled mail mercy missions to Akutan, Cold Bay and Morzhovoi. During summer months, the trip continued on around the Bering Sea coast to the populous settlements of Bristol Bay. "There are supposed to be twelve round-trips made over this RPO yearly, one each month. As a matter of fact there are usually, but not always, as many as eleven actually performed," Thwaites wrote. The round-trip of more than 3,300 miles took from three weeks to four months, observed Thwaites.

A native of the Great Lakes region of Michigan with a string of seafaring ancestors, Thwaites made his way to Alaska in 1905 after years in the RPO from Florida to Washington state. Surviving several train wrecks, he decided that the railroad was too dangerous and asked the superintendent for assignment to ship service. Posted to the "famous little steamer" *Dora*, he signed on as mail clerk for seven years. While Thwaites made his living as a mail clerk, his lasting contribution consists of what he called his hobby of "Kodaking in Alaska," which produced a unique collection of hundreds of photographs of a bustling Southwest Alaska from 1906 to 1914. From his vantage point on the deck of the *Dora*, he chronicled the events, people, places and vessels of the entire coastal region.

Not yet a territory, Alaska was roughly and inefficiently administered by the military. With virtually little monitoring, it was a wide open season on any raw resource that could be mined, trapped, farmed, harpooned, fished or otherwise extracted from the sea and the land. Successive waves of adventurers and migrating Natives created booming settlements where the money was to be made, abandoning them sometimes only a few years later in favor of the next promising lode. By the 1910 decennial census, the *Dora* provided mail, freight and passenger service for more than 2,000 people in more than 30 such communities from Chignik on the Alaska Peninsula to Atka.

"The coming of the mailboat once a month was a great event in the village. It was like a holiday," recalled Edith Newhall Drugg, daughter of the medical missionaries who operated the Jesse Lee Home in Unalaska, a Methodist Church-run home for children. "Schools closed and everyone – villagers with their babies, small children with their dogs – were on the wharf to greet the steamer."

Described by Jesse Lee Home teacher Mary Winchell as "the kindly Mr. Thwaites," the *Dora*'s mail clerk was a popular man. In addition to his mail duties, he often took the time to bring to the people in the far-flung villages small luxuries such as cases of oranges and once, according to Samuel Applegate of Unalaska, a Christmas tree "for the baby." He liked to photograph people. Children and elders, shipmates and travelers, cannery workers and townspeople populate his collection. Most of the images are spontaneous: a newly married couple on the steps of the church, a family with several children in front of their barabara, throngs of people greeting the mailboat at

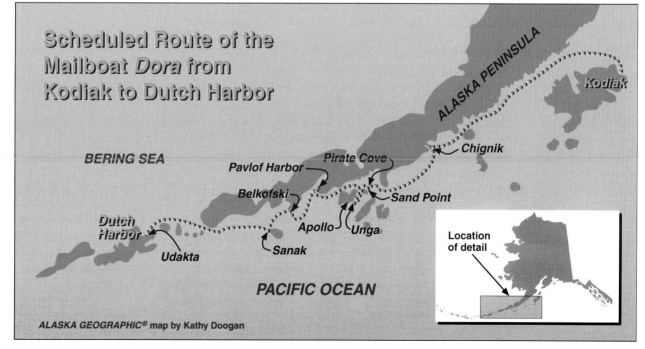

Scheduled Route of the Mailboat *Dora* from Kodiak to Dutch Harbor

BERING SEA

ALASKA PENINSULA

Kodiak

Chignik

Pirate Cove

Pavlof Harbor

Sand Point

Belkofski

Apollo — Unga

Dutch Harbor

Sanak

Udakta

Location of detail

PACIFIC OCEAN

ALASKA GEOGRAPHIC® map by Kathy Doogan

Left: *Born in 1863, John E. Thwaites was raised on a farm in Michigan on the shores of Lake Huron. While his various vocations included schoolteacher, drugstore proprietor and railway mail clerk, he is known for his photographs of coastal Alaska taken aboard the S.S. Dora from 1906 to 1912. On his first trip on the Dora, the ship was adrift with a broken boiler for 63 days. (Alaska State Library, Clyda S. Greely Col., No. 66-842)*

Below: *Sealers Kito Maru, Kaiwo Maru and Nito anchor up under arrest in Iliuliuk Harbor at Unalaska in 1907. The vessels were seized for illegal sealing near the Pribilof Islands by the U.S. Revenue Cutters Bear and Manning. The Manning is shown here tied up at the Unalaska dock. (Alaska State Library, John E. Thwaites Col., No. 18-36)*

the dock, and many snapshots of people traveling aboard the *Dora*, especially children. Married twice but with no children of his own, he seemed to delight in taking photos of children.

A self-taught photographer, Thwaites used a good sense of composition and an artistic eye to shoot his documentary-style photographs. Like his contemporaries, he reproduced his best images on the hottest communication medium of the times, photo postcards. Armed with a Kodak 3-A Special, the Cadillac of the popular postcard cameras, Thwaites most likely constructed a darkroom aboard the *Dora* and there printed hundreds of fashionable cards with exotic Aleutian scenes. To this day, his postcards can be found in many collections.

His photographs covered the era of some of the last seal, sea otter and whale hunts, as well as the frenzied activities of the summer salmon industry and the cod and herring fisheries. On the Alaska Peninsula alone, seven major canneries packed sockeye salmon. More than 80 vessels from tugs to sailing schooners made up the salmon fishing fleet. Sand Point, in the Shumagins, shipped out thousands of pounds of salted cod during those years, and numerous small salteries operated in bays and coves of the Shumagins. By 1913, more than 3,000 people worked in the various fisheries on the Alaska Peninsula.

Quite often, his images depicted the terrors of life in the northern sea-lanes. Despite the treacherous weather, the sea-lanes were busier than ever during the years Thwaites traversed Southwest Alaska. While delivering the mail, the *Dora* often rescued hapless seafarers whose ships were wrecked in storms or foundered on rocks. "Picked up 193 survivors from this wreck [the *Columbia*] and carried them over 1,000 miles. There were 240 of us," the mail clerk wrote to a friend in 1911.

Thwaites photographed more than 45 classic vessels. Among them were Aleut bidarkas, U.S. revenue cutters, Japanese seal poaching schooners, halibut and codfishing schooners, dories, steamships, merchantmen, whalers, tugs, tenders, barges and salmon packers. The revenue cutters pursued the Japanese and American seal poachers. On one occasion when the cutters managed to capture several poachers in 1907, Thwaites, in turn, captured the righteous scene of a cutter guarding the prisoners anchored ignominiously in Iliuliuk Harbor.

In addition to "Kodaking," Thwaites wrote about many of his adventures as a mail clerk. An entertaining storyteller with a sense of wit and equanimity even in life-threatening predicaments, his tales of shipwrecks, volcanic eruptions and other Bering Sea stories were published in several journals of the time. As a special correspondent to the *Seattle Post-Intelligencer*, he wrote dramatic accounts of the wreck of the *Farallon* and the eruption of Mount Katmai. One of his best-known published pieces chronicled his fateful first trip aboard the *Dora*, a trip that became a legend in Pacific Northwest maritime history, when the Christmas 1905 mail arrived at Aleutian villages in spring 1906.

Transferred in 1914 to the Seward-Seattle mail run, Thwaites continued to photograph. He re- tired from the railway mail service in 1919 and settled in Ketchikan where he opened a small photography and curio shop with his second wife, Isabelle Marsters Morse. Thirteen years later, in declining health, he sold his shop. He and Isabelle moved to Mercer Island, Wash., where he passed away in 1940. ◆

Japanese freighters still call at the port where more than a century ago their country's pelagic sealing fleet anchored. This Japanese freighter is anchored in Captain's Bay. (Scott Darsney)

By Lee Dye; photos by Sherie Dye

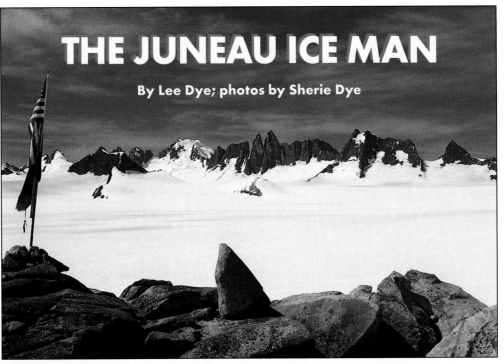

Editor's note: *A resident of Juneau and science columnist for the* Los Angeles Times, *Lee and Sherie, his wife, last reported on the North Slope's Mesa archaeological site for* ALASKA GEOGRAPHIC®.

Our helicopter *whup-whups* up the steep slope of Mount Juneau and like a giant insect with one glass eye peers over the crest of Blackerby Ridge. The pilot and his two passengers strain against their seat belts as if to pull the chopper up over the sharp rocks below, and then collectively catch their breath as they gaze across an other-worldly scene.

The Juneau Icefield sprawls in the distance, ringed by dark spires of rock chiseled by the endless gnawing of the glaciers. The cold landscape is as intimidating as it is spectacular, and though countless others have passed this way, it stretches out before us as if it were virgin land, waiting for the first intruders.

A thin line is etched in the ice below, clearly a human artifact because it is so straight that nature would never have created it in this lavishly sculptured setting. It is a ski trail, the only sign that several dozen scientists, researchers and students have passed this way in recent days.

The helicopter speeds over the trail, covering in minutes what it took the members of the Juneau Icefield Research Project many hours to cross. The upper reaches of Taku Glacier pass quickly beneath us, and off on the horizon several tin buildings shimmer in the summer sunlight, perched on rock outcroppings at the edge of the ice.

The sheds blur past us as the pilot sweeps over the settlement and sets the bird down on the ice. The strong downdraft nearly blows Maynard M. Miller off his feet as he makes his way across the frozen land to greet the chopper. Well past the age when most men have retired, the veteran explorer and distinguished scientist quickly takes charge of his guests. He is the king of this mountain. Barking orders to those who have accompanied him up the steep hill to the landing site, he tells us to wait for him in the cook shack. He will talk with us, he says, when he has finished more important chores.

Heavy rainfall and dense clouds have plagued the expedition, giving way in the last few hours to sunlight so brilliant that it scorches human skin in a matter of minutes.

"When the weather breaks like this we have to move fast and efficiently," Miller says.

This is Maynard Miller's office for two months each summer and it's hard to imagine a more spectacular work place. The Taku Towers rise on the far side of the icefield.

For Miller, the 55 students in his expedition come first, and he wants them to dive into research projects they will have less than two months to complete. He knows, possibly better than anyone else, that the fickle weather on the icefields of Southeast Alaska can change quickly, and there is much work to be done.

Brisk in manner and confident to the point of seeming aloof, Miller hovers over the students as he has now every summer for several decades. He is the founder and lifeblood of a unique summer expeditionary program that has played a crucial role in the training of more than 3,000 scientists.

Those who have completed the program say they will never forget it. And they add they will never forget Maynard M. Miller.

"He's an imposing figure," says Bruce Molnia, director of international polar research programs for the U.S. Geological Survey, who made the trip across the Juneau Icefield with Miller in 1968. "I owe my interest in glaciers to having participated as a student."

Another leading scientist adds that while he has immense respect for him as an educator, Miller is not what he would consider a warm, personal friend. "I have learned how to get around the obnoxious edges," he says confidentially.

Yet this complex man can turn instantly from a stiff field commander to an avuncular, gracious host. Quick with a smile, and surprisingly fit for a man in his 70s, there is a genuineness in his care for his students that transcends all other impressions.

Distinguished in appearance as well as achievement, Miller is a handsome man with a kind face. Despite his snow white hair, he looks years younger than he is, and he could pass for a

The deck of the cook house at Camp 10 is a perfect spot for Maynard Miller to address his students.

banker or stockbroker except for the deep lines in his face carved by the wind and cold of hostile environments.

If he is troubled by the mixed feelings he engenders, Miller does not show it. He knows he has carved his notch in history, and he seems perfectly comfortable with who and what he has become.

Miller was already an experienced mountaineer when he graduated magna cum laude from Harvard University in 1943. He

earned his masters at Columbia University in 1948, and as a Fulbright Scholar he won his Ph.D. in geomorphology and geophysics at Cambridge University, England, in 1957.

If scholarship was his strength, mountains and glaciers were his passion. He came to Alaska as a young man in 1940 to participate in the first ascent of Mount Bertha in the Fairweather Range, and he has been coming back ever since. He led the first American ascent

of the south side of 18,008-foot Mount St. Elias in 1946, and he has participated in expeditions around the world, including serving as chief scientist for the 1963 American expedition up Mount Everest.

His lifelong association with the icefields of Alaska dates back to the 1940s when, as a young Navy officer, he was asked to help determine the feasibility of sending submarines under the arctic ice pack.

"Sea ice was the issue,"

he says. "There was a question of whether sea ice was thickening or thinning."

Even then there was concern among scientists about a subject that still ranks high on their worry list: global change in weather patterns. Today, most scientists think the planet is gradually growing warmer from the "greenhouse effect" caused by an increasing amount of carbon dioxide in the Earth's atmosphere from the burning of fossil fuels. But in those days, scientists thought the planet was heading into a cooling trend. That could thicken the ice cap and eliminate the possibility of sending subs below it.

The ice cap itself is made up of ice so young that it could tell nothing about long-term changes in the weather. To find the answer, scientists needed ice that had been around for more than 100 years.

"The best place to get that was to go to the glaciers," he says.

It takes about a century for the snow that accumulates at higher elevations to compact into blue ice and travel down a glacier to its coastal terminus.

Maynard M. Miller, head of the Juneau Icefield Research Program, awaits the arrival of a helicopter at Camp 10.

The rate of flow, frequently more than a meter in a week, is determined primarily by long-term weather patterns. Miller thinks research carried out the past few decades on the Juneau Icefield has answered the question posed to him so long ago.

The weather, he says, is slowly growing warmer.

If that were all there was to it, our story would end here. But Miller's work has given birth to something that transcends the quest for a single answer.

In 1947, Miller and fellow glaciologist William O. Field established the Juneau Icefield Research Program to study the 38 major glaciers that feed off the icefield in the mountains above Alaska's capital city. After more than a decade of research, a major challenge to U.S. attitudes toward education came from an unlikely source. The Soviet Union launched the first satellite into orbit, and Miller believes that achievement issued a wake up call to American educators.

"The U.S. was stampeded into rapid fire emphasis on science education because the Russians were getting ahead of us," he says.

Miller seized the moment. In 1958, he and his wife, Joan, established the Foundation for Glacier and Environmental Research to raise funds for the program. The funds have come primarily from the National Science Foundation. The University of Idaho, where Miller still teaches, created the Glaciological and Arctic Sciences Institute in 1975, and he has directed that institute ever since.

What grew out of these various programs is a unique project – Miller says it is the only one of its type in North America – that has become the place young scientists go to learn how to lead their own expeditions.

"We're not aiming to develop leaders for expeditions, but that's what happens," Miller says. "We are aiming primarily to develop good scientists for the future. But a good scientist is going to be working in remote areas of the world, and guess what, somebody's got to support them, and those scientists have to become managers too.

"So of course we're training them in that area, but our fundamental goal is not to train leaders, but to train good scientists who must be leaders," he adds.

More than 3,000 students have participated in the program so far, learning about science by carrying out research projects in a most unusual environment. The program lasts two months, July through August of each year, and it challenges both the physical stamina as well as the intellectual capabilities of every participant.

The students must make the entire 150-mile trip up from Juneau and across the 1,700-square-mile icefield to Atlin, British Columbia, on foot and skis. There are 20 camps scattered across the 5,000-square-mile area of rugged peaks and rock outcroppings that encompasses the icefield, which stretches from Skagway in the north to Juneau in the south. The students pause at the camps to carry out their projects. It is a grueling, demanding program.

We had followed them around Juneau for several days as they carried out local field trips. On the day they were to begin their trek up Mount Juneau, they looked beat. Exhausted from studying and hiking and trying to get acquainted, few of them looked as though they could make it. Some of them had never been on skis before.

On the morning they left, unusually heavy rainfall drenched the trail they were to follow, and it was a cold and miserable day to begin such a trip.

We caught up with them several days later by taking a helicopter to Camp 10 on the rim of the icefield. After days of rain and low fog, the sun had finally broken through, giving the students their first glimpse of the great land they had entered.

I was not prepared for the transformation the students had undergone. Far from being tired and exhausted, they were bright eyed and excited, full of life. I asked Jason Mellerstig, a senior geology student at Brown University, what the trip had been like.

"We spent two days hiking through a vertical swamp," he says of the climb up Mount Juneau. "Eight, 10, 12 inches of mud. Just kind of sloshing our way, ducking under logs, jumping over river crossing. Utterly exhausting. You just have to stop sometimes and crash.

"The next day you continue. We were hiking up a 30-degree slope that seemed to go on forever. And then, just barely in the mist, you see an American flag flying (at the first camp.)

"Then we were living in a cloud for four days.

And one morning we woke up and it lifted and we realized we were actually in this incredibly beautiful place. I didn't think it was real. I'm just awestruck."

The interview completed, Jason scurries off to work with the survey crew trying to measure the rate of movement of the ice below us. Other students were busy making meteorological measurements, and preparing to fire explosives to determine the depth of the ice.

The explosives, set off a few feet above the ice, send seismic waves down through the ice, which are reflected back to the surface by the hard rock below. The time it takes for the signals to return tells how deep the ice is. Miller and some of the students completed the third year of that project last summer, and what they have learned surprised even some of the experts.

The experiments revealed that the ice is much deeper than had been thought. Just in front of Camp 10, for example, which is about 19 miles from the coastal terminus of Taku Glacier, the ice is

about 3,800 feet deep, nearly twice the expected depth, and that means that the bottom of the glacier is below sea level. Thus the glacier formed in a deep fiord that had been carved by rivers millions of years earlier.

So when Taku Glacier begins retreating, as Miller expects it to in about a decade, it will leave another spectacular coastal canyon in its wake. With that in mind, several students picked up the tools of their trade and began moving across the icefield, preparing once again to set off explosives and confirm their discovery.

It was going to be a busy day, and with his charges buried in their work, Miller finally has time to talk. He climbs up on a rock above the icefield, looking like the guru he has become. With his bright red jacket shimmering in the sunlight, he has changed from field commander to scholar.

Asked to describe the spectacular setting below

him, he launches into a brief lecture.

"This landscape is the result of 100 million years of geological evolution beginning with the building of the Coast Mountains," he says in earnest professorial tones. "About five million years

ago, the climate began to cool, and it ended up with an ice sheet, with repeated advances over the last two million years. By then, of course, the mountains had been uplifted to a sufficient height that the snowfall didn't all melt away

Members of a survey crew collect their equipment before beginning a trip across the Juneau Icefield.

The survey party in foreground is dwarfed by the wide expanse of upper Taku Glacier. Surveyors perched on ridges overlooking the icefield measure the distance to flags posted by the crew on the ice. As the ice moves, the flags are transported also, thus allowing scientists to determine the rate of flow of the ice.

"This landscape is a written history of that, it is an encyclopedia of that."

It is to this tabernacle of learning that Miller has brought his students, as he has so many years before.

"Instead of bringing nature into the classroom, we bring the classroom and the student into nature," he says.

It is a line he has used many times before, but it still sounds spontaneous. He grows silent for a moment and gazes at the majesty before him.

It is a scene that never grows old.

"This," he says reverently as he waves his arm over the landscape, "is the greatest teacher of all."

The lecture has ended, and Miller bounds off his rock to meet an incoming helicopter. He will relinquish some of his responsibilities soon so that he can devote more time to other challenges. Now a member of the Idaho state legislature, he has other mountains to climb.

Asked if he will continue to come back to the icefield, he scoffs at the question as though only an imbecile would think otherwise. Of course he will be back, year after year, he says.

For Maynard M. Miller, there can be no other choice. ◆

This sign marks the East Tongass National Forest, consisting of one spruce tree growing far above tree line. There used to be two such trees, but a few years ago a student cut one down for a research project, sending her advisors into a state of shock. This episode resulted in the sign: No Clearcutting.

during the year and the glaciers were built up. We have had hundreds and hundreds of advances and retreats of the glaciers.

"There is so much snowfall that it builds up and then flows out to the sea. And it moves out so rapidly that what is forming out here in the last several decades will be down at the terminus of the Taku Glacier 20 miles to the south in about 100 years.

"The glaciers are constantly filing across the land, cutting it and eroding it. So what we are looking at, these high peaks like the Taku Range, very spectacular, sharp serrated spires and horns, were the result of the glaciers cutting against them.

The Mighty Roots of the Aurora

By Carla Helfferich, University of Alaska Geophysical Institute

Editor's note: *Carla recently retired as a science writer for the Geophysical Institute.*

If you've lived for any length of time in the North, then you probably know what causes the northern lights. Properly known as the aurora borealis, these spectacular bands and curtains of light in the night sky can be seen more frequently in Alaska than in any other state, and they shine above the Interior more often than anywhere else. Statistically, Fort Yukon is the aurora capital of the world.

It's understood that the sun ultimately causes the aurora. As geophysicist Dan Swift explained some years ago, "It is almost certain that the energy to power the aurora comes from the sun. From the sun there is a continuous outflow of matter in the form of electrons and nuclei of atoms, mostly hydrogen nuclei (protons). This flow, called the solar wind, streams at speeds near…900,000 mph, and therefore takes several days to reach the earth whereas light takes only eight minutes."

The solar wind powers what amounts to a gigantic electrical generator as it comes streaming in to Earth's magnetic field, and that generator provides the energy and charged particles to turn on the

A green aurora shimmers over the Copper River and the Chugach Mountains in this view from the Million Dollar Bridge east of Cordova. (Cary Anderson)

northern lights. The auroral light itself is emitted from molecules and atoms of Earth's atmospheric gases, in a process similar to the illumination of a neon sign or a television picture tube.

Since Swift's explanation, science has refined its ideas of what causes the aurora. One refinement was to finger solar flares as the cause of auroras and power outages and radio troubles. So what is this solar wind?

Here's where the explanation gets interesting. A couple of years ago, you could have pointed out that solar flares, which are essentially eruptions on the sun's surface, were simply the propulsive mechanism that put gusts in the solar wind. In effect, the flares

serve as the sling, and the protons and electrons are the shot.

But in the past couple of years the experts have changed their tune.

According to the British journal *Nature*, many geophysicists now assert that both auroras and flares are manifestations of upheavals on the sun. The solar surface boils and seethes on a titanic scale, and the magnetic fields embedded within it are incessantly shifting and twisting to match. This great roiling turbulence sometimes brings oppositely directed magnetic fields together. In this so-called "magnetic reconnection," the oppositely pointing fields annihilate one another, releasing X-rays and energetic particles.

Usually the mutually annihilating fields lead to the eruption of a solar flare. Sometimes, however, a certain sort of large-scale magnetic upheaval will produce a far grander event, a coronal mass ejection. Envision a gigantic bubble of sun-stuff, subtending perhaps a quarter of the sun's circumference, bursting out and away from the solar body: that would put quite a kick in the solar wind.

Coronal mass ejections can be observed only from space-borne coronographs, observing instruments that black out the sun's brilliant disk so its outer layers can be studied. Thus, they're fairly recent additions to knowledge of the sun's behavior, and scientists are still figuring out what they do. But they've found some pretty good evidence that their great speed plus the strength or shape of their magnetic fields are what generate large geomagnetic storms and dramatic auroral displays.

Note: *For more information on the aurora, please see Vol. 6, No. 2 of ALASKA GEOGRAPHIC®,* Aurora Borealis.

Steve Zawistowski

By Janet R. Klein

Editor's note: *Janet Klein, a freelance writer living in Homer, has interviewed Steve Zawistowski sporadically since 1981. This article is based on those many interviews.*

Even before he slipped off his overshoes at my cabin door, about five miles east of Homer, Steve Zawistowski called out excitedly, "I just visited a fox farm."

"An abandoned one?" I replied, rising from my desk.

"No!" His voice cracked with excitement as he continued, "Randall Jones just off East End Road is raising foxes. He's got six pairs of silvers."

As Zawistowski relaxed into the rocking chair, the October sun shimmered through the bay window, crossed the carpet, slid up his slacks and warmed his hands. His soft, blue eyes sparkled with excitement yet the sun's warmth, the quiet of the cabin, and

the comfortableness of his memories relaxed him. Only the constant twitching of his thumbs betrayed his inner feelings. As the chair rocked rhythmically, Zawistowski's eyelids drooped, the wrinkles creasing his sloping brow softened, and he shed decades, slipping easily into an Alaska most of us only dream about.

It was 1930 again.

Fresh from herring fishing in the Aleutians, the 23-year-old had just

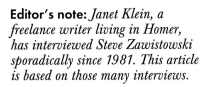

Ever the fisherman, Steve Zawistowski brings skill and long experience to his ice fishing in this 1950 photo. As a commercial fisherman, Zawistowski joined the herring fleet in the Aleutians and for years operated the Normandie during Cook Inlet salmon seasons. (William Wakeland)

purchased interest in a fox farm on Battle Creek near the head of Kachemak Bay. Youthful dreams of financial independence, self-sufficiency and a stable life encouraged investment in the enterprise. He and his partner, H.T. Jansen, entered fur farming with high hopes. Alaska furs were the best in

Steve Zawistowski entered fox farming as it was in decline in 1930. A decade earlier, the industry flourished as exemplified by this unusually neat, compact fox farm on Passage Island near Port Graham at the tip of the Kenai Peninsula. Foxes raised along the coast had coarser fur and a thicker hide which, although acceptable on the market, was slightly less desirable than the finer fur and thinner hide of animals raised in the Interior. (From the collection of Janet R. Klein, gift of S. Zawistowski)

the country and they commanded peak prices in Europe.

Fox farming in Alaska originated when Russian trappers released blue foxes onto isolated Aleutian islands in the 1700s. The feral animals propagated freely and were harvested irregularly. Two centuries later, shortly after World War I, fur prices soared again and revitalized the industry. At a feversome pitch, it spread along the Alaska arc from Southeast to the Aleutians but money, materials and time were needed to achieve success.

Independent individuals settled in remote bays and carved tiny farms in the coastal wilderness. Other than a few trips to town to obtain food and equipment and to ship pelts to market, fox farmers

were self-reliant and resourceful, which appealed to Zawistowski.

Fur farming stirred romantic sentiments. The word "fur" created visions of silky pelts draped across the shoulders of the wealthy; "farming" connoted productiveness and harvesting an annual crop and annual income. It was "...like having a dairy farm, a 24 hour a day job," Zawistowski recalled.

Life revolved around the needs of the foxes and their relationship with the farmer. To house their blues, a favorite color phase of the arctic fox, Zawistowski and Jansen built 20-foot-long wire runs or pens. To "outfox" the canines, they buried the chicken wire deep so the active diggers could not burrow out and stretched it high so the agile animals could not scale it.

As Zawistowski reminisced, it became clear that fox farming was an art, developed with time. So many little things could make a huge difference in this business. Foxes were extremely sensitive to noise, highly responsive to external stimuli, so quiet and routine were important. The remoteness of many farms benefited the business. If exposed to sunshine for even four or five days, the outer guard hairs would discolor and, if noticed by the buyer, the pelt's value would plummet. The plentiful gray days of coastal Alaska also helped the business.

How well Zawistowski remembered hauling in nets heavy with fish, snaring rabbits, occasionally purchasing whale meat from Cook Inlet whalers, digging clams and

mussels in lean months and killing porcupines.

"When we were feeding foxes, we mixed their food. We added a little bit of fur for roughage, an occasional bird and lots of porcupines," he said. "The country was just infested with porcupines. I could walk out onto the tidal flats and where new patches of green grass were appearing, I could find 15 to 20 porkys in a day," he marveled. "I kept a tally one year of the porkys I killed for fox food and it was over 1100."

Obtaining meat was only the first step in feeding the foxes. The meat was then cooked and mixed with imported grains for daily meals or smoked or dried for future use. Feeding was an art. A better diet contributed to a thicker, more luxurious pelt that resulted in a higher price.

Zawistowski and Jansen "pelted off" or killed their animals when their coats were prime, usually in December or January. Only select breeding animals were kept. Attention to detail remained critical for at this time freezing rain often occurred in Kachemak Bay and as Zawistowski recollected, "If a fox sat down, his long guard hairs would freeze to the ground and that didn't bring a very good price."

The furs were boated to Seldovia, to Seattle, to London. Payment, from Lampson and Co., arrived months later. The British pounds were exchanged at the local trading post or bank and the cycle continued.

Although the market was declining at the onset of the Depression, few could predict how rapidly and radically it would

crash. During winter 1931, the Battle Creek fox farmers garnered $100 per pelt. A year later, that same quality pelt brought $11. Zawistowski quit; Jansen struggled for a few more years.

"It took me two years to go broke and I've been scratching for a living ever since," chuckled Zawistowski as his thoughts returned to the present. With good memories lingering, he bid me good-bye.

Steve Zawistowski is a coastal man. His parents, lured west by stories of cowboys and wide open spaces, packed up their three-month-old son and in November 1907 departed New Jersey, crossed the continent by train, and settled in Oregon. The banks of the Columbia River were his playground.

Always, the country captivated Zawistowski. During the summers

of 1933 and 1934, he commercially harvested razor clams at Polly Creek on the west side of Cook Inlet; and, for the next 13 winters, ran a trapline not far from his former fox farm. His wife, Winnie Simpson, maintained their cabin at Clearwater Slough, while deep in the open country at the head of Kachemak Bay Zawistowski trapped lynx, weasels, foxes and coyotes. Coyotes had come into Kachemak country "in droves" in the 1930s and the trapper found porcupine balls everywhere. The crafty canines would flip a porky over, rip the belly skin, then deftly peel it out and around the quills, like turning a glove inside out. The bounty on coyotes was $15 to $20 per pelt; on bald eagles $1 a bird. Forty-two eagles bought the marksman a new rifle one year.

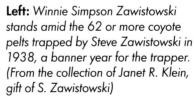

Left: *Winnie Simpson Zawistowski stands amid the 62 or more coyote pelts trapped by Steve Zawistowski in 1938, a banner year for the trapper. (From the collection of Janet R. Klein, gift of S. Zawistowski)*

Right: *Steve Zawistowski holds a prehistoric adze and shares stories of his life in Kachemak Bay at an outing sponsored by the Center for Alaskan Coastal Studies at Kachemak Bay Wilderness Lodge in June 1993. (Doug Loshbaugh, Homer News)*

An unidentified man, left, Steve Zawistowski, center, and his partner, H.T. Jansen, proudly display pelts from blue foxes they raised and from ermine and red foxes (white tip on the tail) they trapped near their Battle Creek fox farm in upper Kachemak Bay. (From the collection of Janet R. Klein, gift from S. Zawistowski)

he chartered his boat to transport gear and people. Some years were good, some not so good.

"Now if you want a good year, every year," he states, "you should be a politician."

His first wife died and in 1983, Zawistowski married Lorna Keeler, herself a pioneer from Anchor Point. With help, they built a comfortable log home east of Homer and, from his windows, he can see the head of Kachemak Bay.

For several seasons, Zawistowski assisted Cecil and Ina Lea Jones, owners of Jones Guide and Outfitting, at their hunting and guiding camp along Sheep Creek, his former trapping territory. He was camp cook, caretaker of the horses and chief storyteller, regaling guests with tales of fox farming, hunting and trapping. After closing camp, the old-timer and his stepdaughter would herd the pack and riding horses back to the Jones Ranch about 18 miles east of Homer. Hip problems and, finally, replacement surgery in 1993 stopped his horseback riding. ◆

In the mid-1930s, the Zawistowskis moved to Seldovia and Steve returned to fishing. Six years earlier, he had fished for late winter herring near Dutch Harbor. En route to Dutch one winter, the schooner he was aboard docked at Seldovia and the young man who grew up in forested Oregon took one look at the spruce and disembarked. He was home.

In 1935, using a set of plans from a New York firm and local lumber, Zawistowski built the 30-foot-long *Normandie*. Until his retirement in 1982 at age 75, he seined and gillnetted salmon in Cook Inlet.

Where did Zawistowski obtain the myriad tools needed for fox farming, trapping and fishing? If you couldn't make it yourself, you went without, or ordered it from Sears and Roebuck. "I just can't see how people could live without Sears and Roebuck. It was your Bible," he says. Even the small, attractive frame house he purchased in Seldovia from Antone Johnson had been a Sears and Roebuck mail-order item.

Zawistowski was hospitalized with tuberculosis in 1942 and 1943 and recovered slowly in the Marine Hospital in Seattle. Upon recovery, he returned to Alaska. Undaunted and undiminished by scarred lungs, he bid on the mail contract and, for four years, boated mail and goods between Seldovia, the Homer Spit and Port Graham. When not running mail or fishing,

ALEUTIAN BROTHERS

Many memorable Alaskans have at one time or another called the Aleutians home. The lives of two of these, step-brothers, mark two milestones in Alaska's marine economy: the end of sea otter hunting and the end of whaling.

Henry Swanson was born Sept. 2, 1895, at Unalaska, the son of a Swedish father who originally settled in the False Pass area and a German-Aleut mother from Unalaska Island. His maternal

Below: *Gen. William Sharrow, assistant adjutant general of the Army National Guard, presents an award to well-known Unalaska resident Henry Swanson at the Unalaska High School gymnasium in June 1982. Seated at the head table at far left is Father Ismail Gromoff of the Russian Orthodox church at Unalaska. At far right stands Philemon Tutiakoff, master of ceremonies. Aang Angagin is Aleut for "Hello, Everybody."(J. Pennelope Goforth)*

Right: *Lighthouse keeper Ed Moore, who referred to himself as one of the last buffalo hunters of the Old West, cuts the hair of assistant keeper Ted Pedersen when Ted returns to Cape Sarichef after his year's leave with pay in 1930. (From the collection of Theodore Pedersen)*

grandfather, Adolph Reinken, managed the Alaska Commercial Co. post at Chernofski. When his father was lost at sea in Unalaska Bay, his mother married a second time to Capt. C. T. Pedersen, famed whaler and sea captain. His stepbrother was Ted Pedersen, keeper of the lonely light at Cape Sarichef on Unimak Island from 1929 to 1933. Cape Sarichef on the Bering Sea side and Scotch Cap on the Pacific side were the first two lighthouses built along Alaska's coast. Scotch Cap remains famous in the annals of marine lore because on April 1, 1946, a 100-foot-high tsunami generated by an undersea earthquake destroyed Scotch Cap lighthouse, killing the five crewman inside.

Henry, who died in 1990, participated in some of the last sea otter hunts in the Aleutians and made his living fox farming in the islands after returning from World War I duty. Ted sailed aboard his father's whaling ship, the *Herman*, and lived through the end of the whaling industry in Alaska waters. When his father later became a trader in furs, he rejoined his dad's ship for one last cruise before that industry too collapsed. Ted later manned lighthouse stations in California before returning to Alaska, to Bear Cove in Kachemak Bay where he died in 1991.

Alaskans and visitors exploring rocky cliffs along the state's extensive, exposed coastline are likely to see the Pacific goose barnacle (Pollicipes polymerus), which lives in the upper two-thirds of the intertidal zone. This species is distributed from the Bering Strait area to Baja California. Goose barnacles generally feed on amphipods and other animals about the size of house flies or smaller. They take their name from a 16th century writer who chanced upon the roots of a beached tree covered with barnacles. The writer took some of the shells back to London, and on opening the shells, found what he took to be birds in various stages of development. (Eric S. Rock)

These stones of pumice, worn smooth by wave action, float in Naknek Lake in Katmai National Park. Pumice is the froth on top of volcanic magma and is composed of primarily gas bubbles. When this foam hits air, it freezes instantly, trapping the bubbles. The density of pumice depends on how much gas it contains; usually the proportion of gas is high, so it floats. Most pumice is composed of siliceous rock, with a chemical makeup roughly equivalent to that of granite. The most common gases found in pumice are water vapor, carbon dioxide, sulfur dioxide and hydrogen sulfide. (Steve McCutcheon)

Index

PHOTOGRAPHERS

ALASKA GEOGRAPHIC. Back Issues

ALL PRICES SUBJECT TO CHANGE

Your $39 membership in The Alaska Geographic Society includes four subsequent issues of ALASKA GEOGRAPHIC®, the Society's official quarterly. Please add $10 per year for non-U.S. memberships.

Additional membership information and free catalog are available on request. Single ALASKA GEOGRAPHIC® back issues are also available. When ordering, please make payments in U.S. funds and add $2 postage/handling per copy for Book Rate; $4 each for Priority Mail. Inquire for non-U.S. postage rates. To order back issues send your check or money order or credit card information (including expiration date and daytime phone number) and volumes desired to:

ALASKA GEOGRAPHIC.

**P.O. Box 93370
Anchorage, AK 99509-3370
Phone (907) 562-0164; Fax (907) 562-0479**

NEXT ISSUE: *Rich Earth: Placer and Hardrock Mining in Alaska*, Vol. 22, No 3. For many, their first impression of Alaska is that of sturdy prospectors trudging across the backcountry and kneeling beside a stream, while visions of wealth swirl in their battered gold pan. After a succession of gold rushes and mining booms and busts, those visions survive. This issue takes a look at the men and women who moiled for gold, the minerals that make up the rich earth and the modern industry that perpetuates Alaska's mining tradition. To members 1995, with index. Price $19.95.